Hybrid Quantum Metaheuristics

Series Page

Quantum Machine Intelligence Series

SERIES EDITORS

Siddhartha Bhattacharyya, Rajnagar Mahavidyalaya, Birbhum, India

Elizabeth C. Behrman, Wichita State University, USA

Hybrid Quantum Metaheuristics
Siddhartha Bhattacharyya, Mario Köppen, Elizabeth Behrman, Ivan Cruz-Aceves

Quantum Machine Intelligence
Siddhartha Bhattacharyya, Mario Köppen, Elizabeth Behrman, Ivan Cruz-Aceves

For more information about this series, please visit: https://www.routledge.com/Quantum-Machine-Intelligence/book-series/QMI

Hybrid Quantum Metaheuristics

Theory and Applications

Edited by
Siddhartha Bhattacharyya
Mario Köppen
Elizabeth Behrman
Ivan Cruz-Aceves

CRC Press
Taylor & Francis Group
Boca Raton London New York

CRC Press is an imprint of the
Taylor & Francis Group, an **informa** business

MATLAB® is a trademark of The MathWorks, Inc. and is used with permission. The MathWorks does not warrant the accuracy of the text or exercises in this book. This book's use or discussion of MATLAB® software or related products does not constitute endorsement or sponsorship by The MathWorks of a particular pedagogical approach or particular use of the MATLAB® software.

First edition published [2022]
by CRC Press
6000 Broken Sound Parkway NW, Suite 300, Boca Raton, FL 33487-2742

and by CRC Press
4 Park Square, Milton Park, Abingdon, Oxon, OX14 4RN

ISBN-13: 978-0-367-75156-2 (hbk)
ISBN-13: 978-1-03-225461-6 (pbk)
ISBN-13: 978-1-00-328329-4 (ebk)

DOI: 10.1201/9781003283294

Typeset in Nimbus font CMR10
by KnowledgeWorks Global Ltd.

Dedication

Siddhartha would like to dedicate this volume to late Prof. John Stewart Bell, FRS, the originator of Bell's theorem, an important theorem in quantum physics regarding hidden variable theories.

Elizabeth would like to dedicate this volume to JPK, EJB, and JFB; and to the memory of CFB.

Mario would like to dedicate this book to the Schrödinger's cat.

Ivan would like to dedicate this volume to his lovely family, his wife Mary, his children Ivan and Yusef, and his mother Bety.

Contents

Editors

Dr. Siddhartha Bhattacharyya received his Bachelors in Physics, and in Optics and Optoelectronics, and Masters in Optics and Optoelectronics from University of Calcutta, India, in 1995, 1998, and 2000, respectively. He completed PhD in Computer Science and Engineering from Jadavpur University, India, in 2008. He is the recipient of the University Gold Medal in Masters from the University of Calcutta. He is the recipient of several coveted awards including the Distinguished HoD Award and Distinguished Professor Award conferred by Computer Society of India, Mumbai Chapter, India in 2017, the Honorary Doctorate Award (D. Litt.) from The University of South America, and the South East Asian Regional Computing Confederation (SEARCC) International Digital Award ICT Educator of the Year in 2017. He has been appointed as the ACM Distinguished Speaker for the tenure 2018–2020. He has been inducted into the People of ACM Hall of Fame by ACM, USA in 2020. He has been appointed as the IEEE Computer Society Distinguished Visitor for the tenure 2021–2023. He has been elected as the full foreign member of the Russian Academy of Natural Sciences. He has been elected a full fellow of the Royal Society for Arts, Manufacturers and Commerce (RSA), London, UK.

He is currently serving as the Principal of Rajnagar Mahavidyalaya, Rajnagar, Birbhum. He served as a Professor in the Department of Computer Science and Engineering of Christ University, Bangalore. He served as the Principal of RCC Institute of Information Technology, Kolkata, India during 2017–2019 and as a Senior Research Scientist in the Faculty of Electrical Engineering and Computer Science of VSB Technical University of Ostrava, Czech Republic (2018–2019). Prior to this, he was the Professor of Information Technology in RCC Institute of Information Technology, Kolkata, India. He served as the Head of the Department from March, 2014 to December, 2016. Prior to this, he was an Associate Professor of Information Technology in RCC Institute of Information Technology, Kolkata, India, from 2011 to 2014. Before that, he served as an Assistant Professor in Department of Computer Science and Information Technology of University Institute of Technology, The University of Burdwan, India from 2005 to 2011. He was a Lecturer in Information Technology of Kalyani Government Engineering College, India during 2001–2005. He is a co-author of 6 books and the co-editor of 80 books. He has more than 300 research publications in international journals and conference proceedings to his credit. He holds 19 patents. He has been the member of the organizing and technical program committees of several national and international conferences. He is the founding Chair of ICCICN 2014, ICRCICN (2015, 2016, 2017, 2018), ISSIP (2017, 2018)

(Kolkata, India). He was the General Chair of several international conferences like WCNSSP 2016 (Chiang Mai, Thailand), ICACCP (2017, 2019) (Sikkim, India), and (ICICC 2018) (New Delhi, India), and ICICC 2019 (Ostrava, Czech Republic).

He is the Associate Editor of several reputed journals including *Applied Soft Computing, IEEE Access, Evolutionary Intelligence, and IET Quantum Communications*. He is the editor of *International Journal of Pattern Recognition Research* and the founding Editor in Chief of *International Journal of Hybrid Intelligence, Inderscience*. He has guest-edited several issues with several international journals. He is serving as the Series Editor of IGI Global Book Series Advances in Information Quality and Management (AIQM), De Gruyter Book Series Frontiers in Computational Intelligence (FCI), CRC Press Book Series(s) Computational Intelligence and Applications & Quantum Machine Intelligence, Wiley Book Series Intelligent Signal and Data Processing, Elsevier Book Series Hybrid Computational Intelligence for Pattern Analysis and Understanding, and Springer Tracts on Human Centered Computing.

His research interests include hybrid intelligence, pattern recognition, multimedia data processing, social networks, and quantum computing.

He is a life fellow of Optical Society of India (OSI), India, life fellow of International Society of Research and Development (ISRD), UK, a fellow of Institution of Engineering and Technology (IET), UK, a fellow of Institute of Electronics and Telecommunication Engineers (IETE), India, and a fellow of Institution of Engineers (IEI), India. He is also a senior member of Institute of Electrical and Electronics Engineers (IEEE), USA, International Institute of Engineering and Technology (IETI), Hong Kong and Association for Computing Machinery (ACM), USA.

He is a life member of Cryptology Research Society of India (CRSI), Computer Society of India (CSI), Indian Society for Technical Education (ISTE), Indian Unit for Pattern Recognition and Artificial Intelligence (IUPRAI), Center for Education Growth and Research (CEGR), Integrated Chambers of Commerce and Industry (ICCI), and Association of Leaders and Industries (ALI). He is a member of Institution of Engineering and Technology (IET), UK, International Rough Set Society, International Association for Engineers (IAENG), Hong Kong, Computer Science Teachers Association (CSTA), USA, International Association of Academicians, Scholars, Scientists and Engineers (IAASSE), USA, Institute of Doctors Engineers and Scientists (IDES), India, The International Society of Service Innovation Professionals (ISSIP), and The Society of Digital Information and Wireless Communications (SDIWC). He is also a certified Chartered Engineer of Institution of Engineers (IEI), India. He is on the Board of Directors of International Institute of Engineering and Technology (IETI), Hong Kong.

Dr. Elizabeth Behrman earned her Bachelor's degree in Mathematics, in 1979, from Brown University. She completed her masters in Chemistry in 1981 and her PhD in Physics in 1985, both at the University of Illinois at Urbana-Champaign. As an undergraduate she was awarded an NSF Undergraduate Research Participation Fellowship and membership in Sigma Xi. At Illinois she was a recipient of a graduate fellowship.

Since 2004, she has been full professor of both Physics and Mathematics at Wichita State University in Wichita, Kansas, where she has earned the Presidents Distinguished Service Award in 2015, and the Academy for Effective Teaching Award in 2012. She was also appointed Kavli Institute for Theoretical Physics Scholar, from 2006 to 2009. Prior to this, she was a professor of Physics from 2002 to 2004, and Chair of the Physics Department from 2003 to 2006. She was an associate director of the honors program at Wichita State from 1999–2003. Prior to this, she was an associate professor of Physics at Wichita State from 1994–2002. During this time she was a recipient of a Lady Davis Fellowship, at Hebrew University, Jerusalem, Israel, 1996–1997. Before that she was an assistant professor at Wichita State from 1990 to 1994. Prior to that, she was an assistant professor of Ceramic Engineering at the New York State College of Ceramics at Alfred, in Alfred, NY. She has published over 60 papers in journals over a wide range of fields, including engineering, physics, and chemistry, reflecting her wide range of interests including chemical kinetics and reaction pathways, ceramic superconductors, glass structure, and nuclear waste vitrification. She was the first to predict the possibility of stable spheroids and tubes made of inorganic materials in 1994. She is also one of the first and major researchers in quantum machine learning. She and her research group have published seminal papers on temporal and spatial quantum backpropagation, quantum Hopfield networks, quantum Bayesian networks, and quantum genetic algorithm.

Mario Köppen studied Physics at the Humboldt-University of Berlin and received his master's degree in Solid State Physics in 1991. Afterwards, he worked as a scientific assistant at the Central Institute for Cybernetics and Information Processing in Berlin and changed his main research interests to image processing and neural networks. From 1992 to 2006, he was working with the Fraunhofer Institute for Production Systems and Design Technology. He continued his works on the industrial applications of image processing, pattern recognition, and soft computing, especially evolutionary computation. During this period, he achieved the doctoral degree at the Technical University Berlin with his thesis works: "Development of an intelligent image processing system by using soft computing" with honors.

He has published more than 150 peer-reviewed papers in conference proceedings, journals, and books. He was active in the organization of various conferences as a chair or member of the program committee, including the WSC online conference series on Soft Computing in Industrial Applications, and the HIS conference series on Hybrid Intelligent Systems. He is founding member of the World Federation of Soft Computing and since 2016 Editor-in-Chief of its Elsevier *Applied Soft Computing* journal. In 2006, he became JSPS fellow at the Kyushu Institute of Technology in Japan, and Professor at the Network Design and Research Center (NDRC) in 2008, and Professor at the Graduate School of Creative Informatics of the Kyushu Institute of Technology in 2013, where he is conducting research in the fields of soft computing, especially for multi-objective and relational optimization, digital convergence, and human-centered computing.

Ivan Cruz-Aceves received a PhD in Electrical Engineering from University of Guanajuato in 2014. He works at the Mexican National Council on Science and Technology (CONACYT) under the project Catedras-CONACYT assigned to the Center for Research in Mathematics (CIMAT) since 2014. He is a member of the Mexican National System of Researchers level 1 and his main research interests are: biomedical signal and image analysis, computational intelligence, and evolutionary computation.

Preface

A metaheuristic is a heuristic (partial search) algorithm that is more or less an efficient optimization algorithm to real-world problems. Hybrid metaheuristics refer to a proper and judicious combination of several other metaheuristics and machine learning algorithms. The hybrid metaheuristics have been found to be more robust and failsafe owing to the complementary character of the individual metaheuristics in the resultant combination. This is primarily due to the fact that the vision of hybridization is to combine different metaheuristics such that each of the combination supplements the other in order to achieve the desired performance. Typical examples use fuzzy-evolutionary, neuro-evolutionary, neuro-fuzzy evolutionary, rough-evolutionary approaches to name a few. Recently, chaos theory has also found wide applications in evolving efficient hybrid metaheuristics.

Quantum computer, as the name suggests, principally works on several quantum physical features. These could be used as an immense alternative to today's apposite computers since they possess faster processing capability (even exponentially) than classical computers. The term quantum computing stems from the synergistic combination of quantum mechanical principles and classical information theory conjoined with principles of computer science. Utilization of the basic features of quantum computing into different evolutionary algorithmic frameworks is foremost part of this research in soft computing discipline. A number of researchers has coupled the underlying principles of quantum computing with various metaheuristic structures to introduce different quantum-inspired algorithmic approaches. The evolution of the quantum computing paradigm has led to the evolution of time efficient and robust hybrid metaheuristics by means of conjoining the principles of quantum mechanics with the conventional metaheuristics, thereby enhancing the real-time performance of the hybrid metaheuristics.

This volume aims to bring together recent advances and trends in methodological approaches, theoretical studies, mathematical and applied techniques related to hybrid quantum metaheuristics, and their applications to engineering problems. The scope of the volume in essence is confined into but not bounded on introducing different novel hybrid quantum metaheuristics for addressing glaring optimization problems ranging from function optimization, data analysis (both discrete and continuous), system optimization, and signal processing to a host of scientific and engineering applications. It is also aimed to emphasize the effectiveness of the proposed approaches over the state-of-the-art existing approaches by means of illustrative examples and real-life case studies.

This volume comprises nine well-versed chapters on different facets of hybrid quantum metaheuristics along with an introductory and concluding chapters. Quantum-inspired metaheuristics can be described as an integrative algorithmic structure, which are designed by exploiting the basics of quantum computing (QC) and metaheuristics. Chapter 1 presents an outline of different categories of quantum-inspired metaheuristics. A brief summary of entanglement-induced optimization is explored in this chapter. This chapter also provides a concise review of W-state encoding-based optimization algorithms.

The performance of any machine learning algorithm depends on the proper choice of parameters of the algorithm and the selected structure of the model. A number of methods have been proposed to solve the problem of optimal choice of a model. These include (i) hyperparametric optimization algorithms and (ii) ensemble methods, in which several machine learning algorithms are used in parallel to collectively solve the problem. Although none of these algorithms guarantee a successful solution to the problem, sufficient experimental evidence is available to prove their effectiveness. As far as classifier decision-making is concerned, the voting method is usually used to collectively combine the decisions of the base classifiers in a machine learning model. In Chapter 2, the authors explore alternative methods of combining such as stacking and ensemble selection, and also propose a new quantum-inspired approach based on metaheuristic hyperparameter tuning algorithms.

Chapter 3 presents the various modern optimization problem-solving techniques available in literature and their limitations. Convergence speed is the primary problem of the modern optimization methods, which has been resolved by using quantum computing. IBM provided quantum computing cloud platform for all the users. IBM Q experience is considered here for field of study. It has two key features, circuit composer and QISkit.

Distribution network (DN) acts as a last mile link between the transmission system and the end users. DN has more power losses in power system network due to the low admittance ratio. Many techniques have been implemented in DN to reduce the losses. Implementation of capacitors with optimal location and capacity, increasing or decreasing the size of the conductors, changing transformer taps, network reconfiguration and optimal allocation of distributed generation (DG) are some of the techniques implemented in DN to reduce the losses. A novel Multipartite Adaptive Quantum-inspired Evolutionary Algorithm (MAQiEA) is proposed in Chapter 4 and used to find the optimal location and capacity of DG. MAQiEA is an updated version of Adaptive Quantum-inspired Evolutionary Algorithm (AQiEA), which is an improvement over QiEA. AQiEA used two Q-bits and entanglement, whereas QiEA had used a single Q-bit. Two modifications have been made in the rotation strategies of AQiEA in MAQiEA. The first modification is made on rotation strategy responsible for exploitation, i.e., rotation away from worse in which bi-partite adaptive crossover operator is augmented with multi-partite operator for better exploitation. The second modification is made on rotation strategy responsible for exploration, i.e., rotation toward better, which has been converted to rotation around better as it has less restriction so provides for better exploration. The effectiveness of MAQiEA is tested on the IEEE benchmark test bus system.

Chapter 5 introduces a simple and effective approach to identify an optimum number of clusters of color images at run. In the real-world problems, determining the appropriate number of clusters from a data set is a challenging task. This chapter proposes a novel automatic clustering algorithm for identifying the optimal number of clusters in color images by using a quantum-inspired framework incorporated with foraging optimization algorithm. The proposed Quantum-Inspired Manta Ray Foraging Optimization (QIMRFO) algorithm has been compared with the classical version of Manta Ray Foraging Optimization (MRFO) algorithm and the well-known Genetic Algorithm (GA) for of color images. The computational results indicate that the proposed QIMRFO outperforms others quantitatively and qualitatively. Furthermore, the utilization of the proposed QIMRFO algorithm can be viewed as an input for solving image segmentation and classification problems. The automatic stenosis detection in X-ray coronary angiograms is very important in medical analysis and information systems. Coronary artery diseases are the most common cause of death in the world over all other cardio vascular diseases.

In Chapter 6, a novel method for coronary stenosis detection using automatic feature extraction is proposed. In a first stage, a set of 31 features were extracted from the image dataset. On a second stage, an automatic feature selection process was performed driven by a Quantum Genetic Algorithm in order to find the most suitable feature subset, able to train a Support Vector Machine-based classifier. The achieved results demonstrate the effectiveness of the proposed method by reducing the number of necessary features for the classification of positive and negative stenosis cases, keeping at the same time an optimal classification rate in terms of the accuracy and the Jaccard index metrics.

Coronary artery disease is the leading cause of mortality rate worldwide. This condition is caused by an abnormal narrowing or occlusion of the artery lumen, known as atherosclerosis (a specific type of stenosis). Currently, many imaging techniques have been tested; nevertheless, conventional X-ray Coronary Angiography (XCA) remains the gold standard for coronary artery disease diagnosis, such as atherosclerosis. Chapter 7 presents a Hybrid Quantum-Convolutional Neural Network to detect coronary artery atherosclerosis in XCA images automatically. It employs a Quantum Convolutional Layer (QCL) that resembles a typical convolutional layer. However, unlike the convolution operation where a classical convolutional filter is applied, a QCL transforms the input using a quantum circuit. This QCL generates a multichannel image from a single channel XCA that feeds a traditional Convolutional Neural Network (CNN). The CNN was trained and optimized using two different architectures previously employed for atherosclerosis detection, viz., a DenseNet-based and a VGG-based architecture. Additionally, a comparison of Stochastic Gradient Descent (SGD) and SGD with Momentum (SGDM) were conducted. These architectures were analyzed and compared using a real dataset composed of 250 real XCA images (split into 125 images for testing and 125 for training) regarding their training without the QCL preprocessing and SGDM. The results showed that the CNNs trained with the outcome of the QCL surpass the baseline CNNs performance.

Automatic clustering of hyperspectral images is a very strenuous task due to the presence of a huge number of redundant bands and complexity to process them. In Chapter 8, two quantum versions of Elephant Herd Optimization algorithm are proposed for this purpose. The use of the binary and ternary quantum logics enhances the exploration and exploitation capability of the elephant herd optimization. These algorithms are compared to their classical counterpart. They are implemented on the Salinas dataset. The proposed qutrit-based algorithm is found to converge faster and produce more robust results. The Xie-Beni Index is used as the fitness function. A few statistical tests like mean, standard deviation, and Kruskal-Wallis test are performed to establish the efficiency of the proposed algorithm. The F score is used to compare the segmented images using the optimal cluster numbers. The proposed algorithms are found to perform better in most of the cases.

Salp Swarm Algorithm (SSA) is a recently introduced metaheuristic algorithm applied for solving benchmark and real-world optimization problems. SSA has good exploitation ability during the search process. However, its exploration ability is limited. Chapter 9 attempts to enhance the balance between exploration and exploitation in SSA by hybridizing with the quantum-inspired framework. The Delta potential-well model (DPWM) from quantum mechanics is known for improving the convergence and diversity in the population to enhance the exploration ability of SSA. The proposed hybrid method is tested using well-known complex convex and nonconvex multiobjective benchmark problems. A comparative study is conducted between proposed MQSSA and well-regarded algorithms MSSA and NSGA-II. The experimental results exemplify that the overall performance of MQSSA is competitive as compared with other approaches.

Chapter 10 proposes a quantum-inspired multiobjective algorithm for automatic clustering of gray scale images. The well-known Non-dominated Sorting Genetic Algorithm II (NSGA-II) inspired by the intrinsic principles of quantum mechanical phenomena is presented here for solving the aforesaid problem by optimizing two objective functions simultaneously. The proposed algorithm has been compared with its classical counterpart and the experimental results over six Berkeley gray scale images certify the efficiency and robustness of the proposed algorithm with reference to the optimal computational time, mean fitness value, standard deviation, and standard error. Finally, a statistical superiority t-test has been conducted between these two algorithms to ascertain the supremacy of the results of the proposed algorithm.

Chapter 11 summarizes the findings reported in the volume with future directions of research.

The book is intended for researchers, academicians, and practitioners. This will serve as a readymade material for the researchers and academicians as it covers a wide range of subject areas belonging to several majors falling under the umbrella of e-waste management. Additionally, it may be treated as a handbook which will suffice the needs of policymakers, supply chain managers, and technology Ninjas. The editors would feel rewarded if the concepts presented in the book come to the social cause.

August, 2021
Siddhartha Bhattacharyya, Birbhum, India
Elizabeth Behrman, Kansas, USA
Mario Köppen, Fukuoka, Japan
Ivan Cruz-Aceves, Guanajuato, Mexico

Contributors

Dora Elisa Alvarado-Carrillo

Department of Computer Science,
  Center for Research in Mathematics
  (CIMAT), A.C.
Jalisco S/N, Col. Valenciana,
  Guanajuato, Mexico

Juan Gabriel Avina-Cervantes

Telematics (CA), Engineering Division
  (DICIS), Campus
  Irapuato-Salamanca
University of Guanajuato, Mexico

Laura-Paulina Badillo-Canchola

Universidad Tecnológica de León. Blvd.
  Universidad Tecnológica No. 225,
  Col. San Carlos, CP. 37670
León, Gto., México

Andrey Batranin

School of Mathematics and Information
  Engineering
Chongqing University of Education,
  China

Siddhartha Bhattacharyya

Rajnagar Mahavidyalaya
Birbhum, India

A Chatterjee

Department of Computer Science
California State University Dominguez
  Hills Carson, CA

D K Chaturvedi

Department of Electrical Engineering
Faculty of Engineering, Dayalbagh
  Educational Institute (Deemed
  University), Agra, India

Debashis De

Maulana Abul Kalam Azad University
  of Technology, Kolkata, India

Sourav De

Department of Computer Science &
  Engineering
Cooch Behar Government Engineering
  College, Cooch Behar, India

Alokananda Dey

Department of Computer Science &
  Engineering
RCC Institute of Information
  Technology Kolkata, India

Sandip Dey

Department of Computer Science
Sukanta Mahavidyalaya, Dhupguri,
  Jalpaiguri West Bengal, India

Tulika Dutta

Department of Computer Science &
  Engineering
Assam University Silchar, Assam, India

Miguel-Angel Gil-Rios

Universidad Tecnológica de León. Blvd.
  Universidad Tecnológica No. 225,
  Col. San Carlos, CP. 37670
León, Gto., México

Sergey Gorbachev

School of Mathematics and Information
  Engineering
Chongqing University of Education,
  China

Soumyajit Goswami
IBM India Private Limited
India

Martha Alicia Hernandez-Gonzalez
Unidad Mdica de Alta Especialidad
(UMAE) - Hospital de
Especialidades No.1. Centro Mdico
Nacional del Bajio, IMSS
Len, Gto., Mexico

Maria-Dolores Juarez-Ramirez
Universidad Tecnológica de León. Blvd.
Universidad Tecnológica No. 225,
Col. San Carlos, CP. 37670
León, Gto., México

Debanjan Konar
SRM University-AP
India

Victor Kuzin
Russian Academy of Engineering
Russia

Ashish Mani
Amity Innovation & Design Centre
Amity School of Engineering &
Technology, Amity University Uttar
Pradesh, Noida, India

G. Manikanta
Department of Electrical & Electronics
Engineering
Amity School of Engineering &
Technology, Amity University Uttar
Pradesh, Noida, India

Emmanuel Ovalle-Magallanes
Telematics (CA), Engineering Division
(DICIS), Campus
Irapuato-Salamanca
University of Guanajuato, Mexico

Sanjai Pathak
Amity University, Uttar Pradesh
Noida, India

Jan Platos
Department of Electrical Engineering &
Computer Science
VSB Technical University of Ostrava,
Czech Republic

María-del-Carmen Ruiz-Robledo
Universidad Tecnológica de León. Blvd.
Universidad Tecnológica No. 225,
Col. San Carlos, CP. 37670
León, Gto., México

Mayank Sharma
Amity University, Uttar Pradesh Noida,
India

Dmytro Shevchuk
School of Mathematics and Information
Engineering
Chongqing University of Education,
China

H P Singh
Department of Electrical & Electronics
Engineering
Amity School of Engineering &
Technology, Amity University Uttar
Pradesh, Noida, India

Vaclav Snasel
Department of Electrical Engineering &
Computer Science
VSB Technical University of Ostrava,
Czech Republic

Sergio Eduardo Solorio-Meza
Universidad Tecnolgica de Mxico
(UNITEC) Campus Len
Len, Gto., Mexico

1 An Introductory Illustration to Hybrid Quantum-Inspired Metaheuristics

1.1 INTRODUCTION

Metaheuristic is basically a stochastic algorithmic framework, which may adopt a predefined set of strategies to build several heuristic algorithms [1][2]. They are efficient, problem-independent approximate algorithms, which can be successfully applied to solve various optimization problems. In the 1970s, the researcher developed a new kind of algorithm, which integrates fundamental heuristics in higher level frameworks to effectively explore a search space. These techniques are nowadays popularly known as metaheuristics. There exists a number of metaheuristics in the literature, which may include, but not restricted to, Particle Swarm Optimization (PSO) [3], Ant Colony Optimization (ACO) [4], Simulated Annealing (SA) [5], Tabu Search (TS) [6][7][8], Differential Evolution (DE) [9] to name a few.

Optimization is an effort of attaining the best possible outcomes under a specified state of affairs. In many real-life scenarios, such as maintenance, construction, and other several cases, a concrete decision may need to be taken to reach at certain goal. Generally, the outcome of such problems is achieving either minimum effort or maximum benefit. This effort can be typically manifested as a function of a set of specific variables [10][11]. As a consequence, optimization can be coined as the process of deriving the criteria, which provides the minimum or value merit of a function. When the optimization deals with a single objective function, it is called "single objective optimization." In case of "multi-objective optimization," there must have a systematic approach to be followed to derive the best possible solutions by using at least two objective functions simultaneously. In some occasions, the solution set is called pareto-optimal solutions in multi-objective optimization [12][13].

The paper is arranged in the following manner: The outline of different quantum-inspired metaheuristics is briefly presented in Section 1.2. This section presents the detailed discussion on metaheuristics of different types. Section 1.3 provides few important entanglement-induced optimization techniques that are available in the literature. The W-state encoding of various optimization techniques is presented and discussed in Section 1.4. Section 1.5 throws light on several quantum system-based optimization techniques. Different bi-level and multi-level quantum-based systems have been highlighted in this section. A few popular applications of quantum-inspired metaheuristics are presented in Section 1.6. The chapter finally concludes with relevant information in Section 1.7.

DOI: 10.1201/9781003283294-1

1.2   QUANTUM-INSPIRED METAHEURISTICS

In essence, the metaheuristics are characterized by high-level strategies to explore search spaces by utilizing different techniques. The "diversification" and "intensification" are two popular terms used to define functioning of metaheuristics. The first and second terms generally point to the exploration of search space and the search experience gained from diversification. There is a balance between these two to execute metaheuristic application. This helps in speedy identification of regions in the search space, which possesses high-quality solutions. In addition, it helps to identify the useful regions of the specified search space in a very small time frame that have already been explored and also the regions, which cannot furnish any high-quality solutions.

The metaheuristic algorithms can be classified and described in different ways based on a specific point of view. This classification is shown in the following points:

1. Nature-inspired metaheuristics vs. non-nature-inspired metaheuristics
2. Memory-based methods vs. memory-less methods
3. Single point-based search vs. population-based search

The metaheuristics as shown in point 1 is basically done by considering the origins of various algorithms. Representative of nature-inspired algorithms may include Fuzzy Systems (FS), Artificial Neural Networks (ANN), Evolutionary Algorithm (EA), and Swarm Optimization (SO), etc. So far, these algorithms have successfully been applied for solving different real-world problems. Particle Swarm Optimization (PSO), Ant Colony Optimization (ACO) [9] belongs to SO, whereas Genetic Algorithm (GA) and Differential Evolution (DE) are two examples of EA. Some popular non-nature-inspired metaheuristics are known to be Tabu Search (TS) and Iterated Local Search (ILS). The metaheuristics are often classified based on the memory utilization, i.e., whether they deploy memory or not. Memory-based metaheuristics as referred to point 2 generally use memory at every iterations and save the search history. On the contrary, memory-less algorithms perform other tasks, for example, sometimes they might accomplish a Markov process, and based on the information gained, the next course of action is determined in the search process. Lastly, metaheuristics of point 3 are classified into two types, called a single point-based metaheuristics and population-based metaheuristics.

In general, algorithms that use a single solution at any point of time are called trajectory methods. These methods include such metaheuristics that deal with local search, like TS, ILS, and variable neighborhood search (VNS). In these metaheuristics, the search procedure expounds a trajectory in the said search space. On the other hand, other one (population-based metaheuristics) can be expounded as the evolution of either points or a probability distribution in/over the search space.

The quantum-inspired metaheuristics are developed by considering two different subareas of computer science called quantum computing (QC) and Evolutionary Computing (EC). Unlike "pure" quantum algorithms, such as Grover's search algorithm [14] or Shor's factorization algorithm [15], in which the functional quantum computer is necessitated for their efficient execution, the quantum-inspired algorithms exploit the features inspired by theory and principles of quantum mechanical systems, like qubits (quantum bits), superposition of states, etc. It necessitates these ideas for developing a computing paradigm much speedy than the conventional computing framework. The viable reason for this enhanced computing speed is achieved by dint of exploiting the inherent parallelism perceived in the qubits, the basic unit of a quantum computer. This makes QCs to be far more effective in comparison with their classical counterparts for obtaining factorization of large numbers [15] and searching elements from databases [14].

Quantum-Inspired Metaheuristics are comparatively the new area of research of metaheuristics class. They draw their faces fundamentally from two dissimilar fields, viz., Metaheuristic and Quantum Computing. Quantum computing is fundamentally harnessing and utilizing the astounding laws of quantum mechanics for processing information. A conventional computer (traditional computer) employs long stream of "bits," which essentially encode either 0 or 1. On the contrary, a quantum computer employs quantum bits or qubits. A quantum bits is a quantum system, which basically encodes 0 and 1 into two perceptible quantum states. Owing to the fact that quantum bits behave quantumly, the happening of "superposition" and "entanglement" can be capitalized for getting better efficiency. A quantum computer is able to process a huge number of calculations at once. Unlike a classical computer, which works with 0s and 1s, a quantum computer can use 0s, 1s, and their superposed form. Hence, using the amazing features of quantum computing, called "superposition" and "entanglement," a quantum computer can perform any certain impenetrable tasks more efficiently and more quickly in comparison with their classical counterpart.

1.2.1   LOCAL SEARCH METAHEURISTICS

Nowadays, local search metaheuristics gained very much popularity among other different approaches. Local search (LS) makes an effort to find good solutions in an iterative way by changing the current solution. The said changes are generally called moves. These moves are usually "small" keeping it in mind that the adjacent solutions are comparatively in close proximity to each other.

The neighborhood of a solution can be described as the set of solutions, which are achieved due to a single move to the specific solution. The general and most common approach adopted by these algorithms is that instead of taking the best improvement solution, they explore a predefined neighborhood to find the best move. They usually do not explore neighborhood at a whole; rather the first improving move is espoused by these kinds of algorithms by applying their first improvement strategy. The move can be varied for different occasions and can be defined accordingly. At each iteration, a solution is being chosen from its neighborhood to replace the current solution using a specified rule called search strategy.

The local search metaheuristics are generally found to be computationally effective since in most cases it takes lesser time to search for an improving move. There exist numerous search strategies in the literature; out of them, steepest ascent or steepest descent is probably the most common one. In this strategy, the best move is found from its neighborhood and used for its later operation. This strategy is often used in metaheuristics that is called hill-climbers. The other popular strategy is popularly known as mildest ascent/descent strategy, which usually chooses that particular solution that improves the present solution by a very little amount. On the other hand, one important move strategy may be exemplified as the first improving strategy, where instead of other moves, the first move is chosen that improves the present solution.

Simulated Annealing (SA) is a very popular metaheuristics, which employs a move strategy that imitates the annealing procedure of a crystalline solid. It was first introduced by Kirkpatrick *et al.* [5] in 1983, which was designed on the basis of the method as presented by Metropolis *et al.* [16] in 1953. A concept called "local optimum" was introduced to signify the solution, which is better than any other solutions in the neighborhood. The "global optima" is the best solution to be found in any optimization problem. When it is found that the current solution falls into a local optimum, there must have a strategy that the metaheuristic must adopt to "escape" from this situation. These kindx of metaheuristics are basically structures that depend mainly on iterative improvement for finding good solutions.

Other popular and commonly used strategies that are very frequently espoused to have a random change, called perturbation to the present solution. Two among them are popularly known as Iterated Local Search (ILS) and Multi-start Local Search (MLS), respectively [17].

One more important strategy of this category uses memory structures to record information of the previous progresses of the search procedures aiming to acquire good solutions. The commonly used metaheuristics that adopt this strategy are termed as Tabu Search (TS) [6][7][8] algorithms. Different categories of memory structures can be used to recollect specific features of the trajectory that the algorithm has taken up in its search space. A tabu list stores the last solutions that has been encountered during operation and prohibits these solutions such that they are not visited again until they are available on the list. Other popular kind of metaheuristic, known as Guided Local Search (GLS) [18], presents a dissimilar type of memory, known to be an augmented objective function, which encompasses a penalty factor for each of the potential element. While it reaches a local optimum, the penalty factor is increased for each element of the present solution. This allows the search process to escape from falling in the local optimum.

1.2.2 CONSTRUCTIVE METAHEURISTICS

Constructive metaheuristics represent a discrete class from the perspective of local search metaheuristics in such a way that they usually contrive solutions from their basic elements, rather operating on complete solutions. In their way, they start operating taking an empty set and add a single element in the course of each iteration. It can be noted that an operation can also be referred to as a move. Like other metaheuristics, they also continue their execution until a solution is found out or other predefined stopping criteria are met.

There exists few popular constructive metaheuristics in the literature, out of them Ant Colony optimization and Greedy Randomized Adaptive Search Procedure (GRASP) are the most popular and mostly used metaheuristics. The basics of these two methods are described below.

Foraging behavior of real ants is the sole inspiration for designing a very popular and well-admired metaheuristic approach, called Ant Colony Optimization [19][20]. This behavior empowers ants to discover shortest paths between their nest and the food sources. In their way, ants try to explore the surroundings of their nest at random. Once an ant succeeds to locate a food source, thereafter carries some food to their nest. While they are on the way of returning to nest, the ant spurts a chemical from its body, called pheromone on the ground. More ants traverse the same path means more amount of pheromone deposited, which in turn attract other ants to follow the same path for food source. Hence, this chemical (pheromone trails) empowers the real ants to communicate with each other indirectly so that they can easily locate shortest paths to reach at food sources from their nest. This amazing behavior of real ant colonies is capitalized on artificial ant colonies for solving different optimization problems. Ant Colony Optimization (ACO) is a very popular example to design meta-heuristic algorithms that can effectively be used to solve various Combinatorial Optimization (CO) problems. In 1991, Dorigo *et al.* [19, 20] first presented an algorithm within this framework. Since then, several algorithms, especially different variants of this basic algorithm, have been presented in the literature. In ACO, a parameterized probabilistic model, sometimes known to be the pheromone model, is imposed to define simulation of this chemical. This model comprises a set of parameters, whose values behave like a memory that can be used to keep track of the search procedure. Several quantum-inspired ant colony optimization techniques have been introduced in the recent years. Wang *et al.* [21] developed a novel evolutionary optimization technique where qubit was represented in ACO. Dey *et al.* [22][23][24] developed few popular quantum-inspired metaheuristics that have been designed for bi-level/multi-level thresholding of gray scale/color images.

Like ACO, GRASP is also a popular metaheuristic use to solve variant problems of combinatorial optimization. GRASP constructs a greedy randomized solution and triees to improve it through a local search in succession within different iterations [25]. The GRASP basically exploits in two phases, called construction phase and local search phase. The greedy solutions (randomized) are usually generated by appending elements to the set of solution of the given problem. They are chosen from the list of elements, which are basically ranked by a selected greedy function. Each solution is ranked on the basis of its quality which it will attain. The elements having well-ranked are frequently put down in a restricted candidate list (RCL) for acquiring variability in the said candidate solutions. The solution in RCL is chosen at random at the time of building up the solution. Hart and Shogan [26] first introduced this greedy randomized construction approach in 1987; sometimes this method is called semi-greedy heuristic.

In the literature, there exist several variations of this algorithm, like the Reactive GRASP. In Reactive GRASP, the parameter that is used to define the limit of the RCL during its first phase is self-adjusted based on the quality of the solutions found earlier [27]. In addition, there also exist several other techniques of this kind; some of them are employed to speed up in searching in various fields, like cost perturbations, memorization and learning, and many others [28].

1.2.3 POPULATION-BASED METAHEURISTICS

These metaheuristics deal with a population comprising a set of solutions at each iteration instead of a single solution. The individuals in the population are updated at every iteration using certain operators provided by the particular metaheuristics. This is essential for getting population diversity, aiming to have better outcome at every iteration. Population-based metaheuristics explore the search space in a consistent, intrinsic way. The performance of different metaheuristics differs, mainly because of the way of manipulation of the population.

Population-based metaheuristics are intended to find favorable solutions by iteratively choosing and thereafter combining current solutions from a set, known as population. The paramount members in this class are popularly known as evolutionary algorithms as they imitate the features of natural evolution. The term "EA" can be used to present batch of algorithms that surrounds the extensive range of evolution-based metaheuristics. Some representative of most investigated population-based metaheuristics are Genetic Algorithm (GA) [29][11], Evolution Strategies (ES) [30], Evolutionary Computation (EC) [31], Differential Evolution [9] to name a few. The population of individuals is updated differently for different population-based methods. For example, the recombination and mutation operators are used to modify the population of Evolutionary Computation, and the same is occurred for ant colony optimization by the guidance the pheromone trails and other heuristic information. Different population-based metaheuristics are described in the following subsections.

Deterministic population-based algorithms for EA are known to be Path Relinking (PT) and Scatter Search (SS) [32][33]. SS encodes solutions as real-valued vectors and thereafter locates new solutions by producing concave or convex linear combinations of the said vectors. PT, on the other hand, introduces the idea of a path between various high-quality solutions, basically generalizing the idea of linear combination. Paths generally comprise elementary moves, which may be used by one in local search metaheuristics. This moves to a path that transmute a single solution (initiating solution) into a corresponding second solution (guiding solution) one at ones. PT can therefore be appraised a local search heuristic, where the move to be carried out is selected on the basis of the condition that the move must guide the solution in close proximity to the guiding solution. The population (also called reference set), which is used usually used to select initiating and guiding solutions, is also updated after the generation of the new solutions as stated by deterministic rules in both SS and PT.

1.2.4 HYBRID METAHEURISTICS

In the literature, metaheuristics have been exhaustively and extensively applied in solving different global optimization problems [34][35][36]. In the recent years, the idea of coalescing several metaheuristics has come out, that is popularly known as hybridization of metaheuristics or, in simple word, hybrid metaheuristics. The hybrid metaheuristics are intended to exploit the features of different metaheuristics and their optimization strategies [37]. The concept of hybridization can play a very effective role, especially in solving different complex optimization problems. The selection of right combination of metaheuristics usually instigates in achieving optimum performance. The number of researchers has presented several hybrid systems in the literature that have been efficiently and successfully applied in solving traditional optimization problems [38][39][40][41][42]. Two popular hybrid architectures of this are Multi-agent Metaheuristic Hybridization (MAMH) [38] and Multi-agent Metaheuristic Architecture (MAGMA) [43].

1.3 ENTANGLEMENT-INDUCED OPTIMIZATION

Quantum entanglement is an important and interesting concept in quantum computing. It is somehow related to the observed physical experience that happens at the time of generating, communicating, and sharing spatial proximity of a set of two or group of particles in such a way that the quantum state of the said particles cannot be elucidated separately of the state of the others, in spite of the particles being kept in separation at a large distance. The term "optimization" is referred to the method that can be used for successfully locating the best solution among the obtained feasible solutions. There are different kinds of optimization problems (continuous or discrete) available that can be grouped into a pair of categories based on the variables used to define the problem.

Of late, several quantum-inspired evolutionary algorithms (QIEAs) have been introduced in the literature. These algorithms generally have expertise in their search ability maintaining a healthy balance between exploration and exploitation. QIEAs can usually be considered as a type of estimation of distribution algorithms (EDAs) [44][45][46]. Han and Kim first presented this kind of algorithm (QEA) in 2002 [47]. This algorithm has the ability to [47] [48] explore a search space using very little amount of individuals and exploit efficiently to find the global solution. A lot of researchers have coalesced QEA with other popular heuristics, such as cuckoo search algorithm, PSO, gravitational search, immune clonal algorithm, and tabu search to propose various quantum-inspired algorithms. The principles of QEA are combined with these quantum-inspired approaches to form several quantum-inspired algorithms. Despite having certain drawbacks, QIEAs may possess several appealing features, which may include better search abilities, least computational costs, and even simple from implementation point of view in comparison with other existing metaheuristics.

The entanglement induced optimization algorithms are new achievement from research point of view. Like other quantum-inspired optimization algorithms that exploit various features of quantum computing in the development process, the entanglement induced optimization algorithms use the quantum entanglement feature to develop these kinds of algorithms. When dealing with high-dependence problems, the conventional optimization techniques have been proved to be non-efficient, especially in locating the global optimum. This problem has been addressed by introducing a novel meta-heuristic, called the entanglement-enhanced quantum-inspired tabu search algorithm (Entanglement-QTS) [49].

The quantum-inspired tabu search (QTS) is a popular, effective, simple, and robust meta-heuristic. Unlike other QIEAs, this novel (QTS) algorithm interestingly applies both the best and worst solutions. The reason behind this is to guide the individual to chase toward finding a better solution and also to go out of the way from a worse solution. QTS can show its efficiency to speedily reach at the global optimum. The several applications of QTS algorithm can be found in the literature [50][51][52], which have encouraging and incomparable searching capability. Entanglement-QTS uses backbone of the quantum-inspired tabu search and the encouraging feature of quantum computing, called quantum entanglement.

In comparison with the other QIEAs, the Entanglement-QTS uses qubits, which are in entangled states. Basically, these entangled states can be demonstrate as an inflated degree of correlation, render the variables twine together. It signifies a state-of-the-art thought that can notably perk up to deal with high-dependence and multimodal problems. This algorithm is able to find optimal solutions, balance in diversification and intensification. Entanglement-QTS uses quantum not gate, which helps to escape several local optimal solutions, strengthen the intensification consequence by entanglement local search, and speed up the optimization procedure by applying entangled states.

1.4   W-STATE ENCODING OF OPTIMIZATION ALGORITHMS

The W state is a very new, interesting, and appealing feature of quantum computing. This can be defined as an entangled quantum state comprising three qubits as presented below.

$$|W_s\rangle = \frac{1}{\sqrt{3}}(|001\rangle + |010\rangle + |100\rangle) \qquad (1.1)$$

where W_s represents the said quantum entangled state.

This is an astonishing property of quantum computing that is usually used to represent a certain class of multipartite entanglement. It may be very useful for taking part for performing different applications in several fields like quantum information theory. The W_s can be thought as the representative of one non-bi-separable class comprising three-qubit states. The interesting feature is that it can not be changed by any means into each other using any of the local quantum functioning [53]. One interesting characteristic of W_s state is that it is robust as because when one qubit is lost by any means, the entanglement property among the participating states are not getting destroyed as a whole; rather, the other two states will be remained entangled to each other.

The entanglement feature is an encouraging topic in the field of research now adays. Many researchers have found interest in working in this direction. In optimization purpose, several researchers have widely used quantum entanglement in a variety of ways [49]. The computation power of W_s state has been elaborately described in [54]. W_s state has few appealing application in communication point of view. A protocol, called quantum secure direct communication (QSDC), has been introduced by Chen et al. [55] in 2008. The foundation and applications of W state have been described in [56]. Cruz et al. presented few W_s state-based efficient quantum algorithms in 2019 [57]. Like other encoding schemes, W-state can be encoded to form population in different quantum behaved applications. The appealing feature of W-state, in turn, can assist to form efficient algorithms, especially in optimization perspective.

1.5 QUANTUM SYSTEM-BASED OPTIMIZATION

Optimization problem deals with observing the best possible solutions to any criteria-based problem. In general, any optimization problem is expressed as either a maximization problem or a minimization problem. There are several optimization techniques available in the literature that can be successfully applied in numerous fields like engineering, economics and mechanics, etc. The potential of quantum computing may encourage to solve problems, which may not realistically be feasible using traditional computers. In addition to this, there may have a considerable speed-up in execution with regard to the popular classical algorithm. The appealing features of quantum computing might help to build quantum optimization algorithms that are built to solve optimization problems. There are several algorithms available in the literature that have been basically designed for dealing with several optimizations with regard to image processing [23]. The following subsections illustrate the bi-level and multi-level quantum system-based optimization.

1.5.1 BI-LEVEL QUANTUM SYSTEM-BASED OPTIMIZATION

Many researchers have already designed several quantum behaved system for bi-level optimization. MirHassani et al. presented a Quantum Binary Particle Swarm Optimization (QBPSO) algorithm [58] in which an effectual nested strategy was adopted to deal with competitive facility location problems. This problem is about to apprehend the almost all of a given market, to make maximum profit. Zhang et al. [59] presented a novel strategic bidding optimization method using bi-level programming and swarm intelligence. This proposed method works in two steps. Firstly, the idea of generalized Nash equilibrium has been used to develop a general multi-leader-one-follower nonlinear bi-level (MLNB) optimization method. Afterwards, a PSO-based method has been developed for solving the problem expounded in the MLNB decision model. Dey et al. [60] introduced quantum behaved bi-level optimization method for image thresholding. The authors proposed two novel algorithms, called quantum-inspired genetic algorithm and quantum-inspired particle swarm optimization to find optimum threshold value from grey scale images. Zhang *et al.* [61] proposed an elite quantum-inspired PSO (EQPSO) algorithm, where an elite approach is taken for the best particle (global best) to avert premature convergence of the swarm. This algorithm has been employed to solve bi-level multi-objective programming problem. The working principles of MRLDE have been studied by Kumar *et al.* [62]. The authors have developed a recently introduced variant of DE merged with Otsu method. Chang *et al.* [63] presented a bi-level semantic representation analyzing approach that can eliminate the shortcomings of existing systems. Yan *et al.* [64] developed a bi-level programming model that can be applied to optimize the strategy, taken for augmentation of transportation network capacity within the proposed budget. The lower and upper level problems face some difficulties, which were solved by proposing a quantum-inspired evolutionary algorithm in this paper. Later, the quantum behaved procedure relating to bi-level image thresholding has been proposed using simulated annealing by Dey *et al.* [65].

1.5.2 MULTI-LEVEL QUANTUM SYSTEM-BASED OPTIMIZATION

Like bi-level quantum behaved system, there exist many applications in the multi-level domain. The functionality of a bi-level system can be enhanced to its corresponding multi-level frame by modifying its computational structure. The complexity, in general, may increase for this required modification. Dey *et al.* [66] presented two quantum-inspired approaches, namely, Quantum-Inspired Simulated Annealing and Quantum-Inspired Ant Colony Optimization that deal with multi-level thresholding. A novel evolutionary algorithm has been presented by Tkachuk [67] that can efficiently handle optimization problems. This algorithm has been built using quantum computations technology. In this algorithm, the notion of many-valued quantum logic has been incorporated, which in turn made the proposed system more efficient than other classical quantum genetic algorithm. Dey *et al.* [22][23][24] proposed several quantum-inspired metaheuristics for color image thresholding in multi-level domain. A protocol for performing quantum reinforcement learning (QRL) with quantum technologies (QTs) has been presented by Cárdenas *et al.* [68]. In comparison with recent QRL, this method does not rely on coherent feedback during its learning process, which in turn enables its implementation in an extensive diversity of quantum systems. Niemann *et al.* [69] proposed a strategy that has the ability to verify the equivalence of two different quantum behaviors irrespective of the dimension of the specified underlying quantum system. This proposed system can be assimilated into data-structures like Quantum Multiple-Valued Decision Diagrams (QMDDs), which enables an effectual verification of the proposed system. Carrasco *et al.* [70] presented a paper, where they investigated a fast and sturdy approach to control the quantum states of a given multi-level quantum system. For this purpose, they used a dual frequency time-varying potential. Grace *et al.* [71] presented a formalism to encode the logical basis of a qubit into multiple physical level's subspaces. Roy and Das [72] presented a multi-level quantum system, where optimal control of this system has been synthesized. The steering problem associated with this quantum system has been solved by minimizing energy cost functional of this control.

1.6 APPLICATIONS OF QUANTUM-INSPIRED METAHEURISTICS

Quantum-inspired Particle Swarm Optimization (QPSO) algorithm [73] presented by Gao *et al.* with the help of quantum evolutionary theory and PSO. This algorithm is employed to optimize some benchmark functions and also some cognitive radio spectrum allocation problem. A quantum-inspired evolutionary algorithm (QEA), named quantum-inspired Tabu search (QTS), is presented in [74]. The classical Tabu search algorithm and the characteristics of quantum computation, such as superposition, are applied to form this method. This proposed method applied to solve different types of NP-complete problems. A new metaheuristic algorithm, named the entanglement-enhanced quantum-inspired tabu search algorithm (Entanglement-QTS) proposed by Kuo and Chou [75] to take care of combinatorial and numerical improvement issues. This algorithm depends on the quantum-inspired tabu search (QTS) algorithm in blend with the quantum entanglement. The authors gave emphasis on the entangled Q-bits, which can extemporize the treatment of multimodal and high-dependence problems significantly. Wang *et al.* [76] applied an improved version of the conventional quantum genetic algorithm to solve different types of optimization problems. This approach can decide the rotating angle, the self-adaptive rotating angle strategy likewise with the quantum disaster operation, and quantum mutation operation. A survey and analysis of the quantum-inspired evolutionary algorithm is illustrated by Xiong *et al.* [77]. Among different operators for the operation of quantum gate update in quantum algorithm, Quantum rotation gate (QRG) gave emphasis in this article. From the start, an itemized investigation on the classification of QRG performed with the assistance of rotation direction and the magnitude of rotation angle by analyzing and condensing different sorts of QRGs and from that point forward, the comparing definitions, depictions, and analyzes are delineated [77]. A comparative study also presented with the help of different kinds of complex function optimization problems. The hybrid quantum-inspired firefly and particle swarm optimization (QIFAPSO) algorithm presented in [78] to solve continuous optimization problems. In this method, the basic concepts of quantum computing, i.e., superposition states of Q-bit and quantum measure used to obtain better diversified solutions. The firefly algorithm and PSO algorithm are adapted with the quantum representation for potential solutions.

Quantum-inspired metaheuristic algorithms are also applied for pattern recognition and image analysis. Quantum-inspired Particle Swarm Optimization (QiPSO) in combination with Evolving Spiking Neural Network (ESNN) applied for string pattern recognition [79]. A quantum genetic algorithm (IQGA) with the help of adaptive adjustment strategy of the rotation angle and the cooperative learning scheme is presented for multilevel thresholding-based image segmentation [80]. In this article, the adaptive adjustment strategy tried to improve the convergence strategy, search ability and stability, and the cooperative learning improved the search ability in the high-dimensional solution space. Dey *et al.* [81] presented a quantum-inspired sperm whale metaheuristic algorithm to multilevel thresholding-based image segmentation. The fundamental operators of quantum computing fused with the sperm whale metaheuristic algorithm. Color MRI image segmented efficiently by the proposed quantum-inspired genetic algorithm-based FCM [82][83]. Liu *et al.* [84] presented a chaotic quantum-behaved PSO based on lateral inhibition (LI-CQPSO) algorithm to solve the image matching problems. In this approach, the chaos theory, and quantum and lateral inhibition combined to get better result. The problem of local best of PSO can be solved by chaos, the searching performance, as well as parameter control of PSO can be handled efficiently by quantum and the edge of the images can be extracted by the literal inhibition. The fuzzy C-means (FCM) in combination with four-chain quantum bee colony optimization (FQABC) applied to segment the images [85]. The four chains quantum encoding method induced in the artificial bee colony (ABC) optimization algorithm, i.e., QABC, to improve the performance of ABC algorithm and the improved QABC algorithm is employed to search the optimal initial centers of FCM.

Swarm intelligence also made a notable signature in the field of robotics. Due to self-autonomy and unexampled levels of distributed intelligence, advancement in the field of robotics is notable. Robots can be applied in different real-world scenarios, like complex manufacturing, health monitoring, disaster management, complex logistics, etc. To make it more effective, the swarm intelligence, a significant part of the computational intelligence, can be merged with the robotics. Combination of these two fields is known as swarm robotics. Swarm robotics successfully applied for smart sensing, correspondence and association functionalities blessed to these small robots, which take into consideration cooperative data detecting, activity and information deduction from the environment [86]. Osaba *et al.* [86] illustrated an overview of the research contribution in the field of swarm robotics, which will end up being a significant research impetus of the Computational Intelligence arena in the recent future. Tan and Zheng [87] narrated a detailed survey in the field of swarm robotics. Basically, inspired from the nature, swarm robotics is a combination of the vast field of swarm intelligence and robotics. Contreras-Cruz *et al.* [88] suggested a hybrid algorithm of artificial bee colony and evolutionary programming to solve the mobile robot path planning problem.

In the field of optimization engineering, real-time tasks scheduling in different complex environments is an all-time daunting task for the researchers. Some of the relevant and prominent research areas of real-time tasks are manufacturing automation, telecommunication systems, multimedia applications, robotics,control system, embedded systems, etc. As the time progresses, the applications of real-time tasks are becoming more complex, time-consuming, and sophisticated. The application area of quantum-inspired genetic algorithm (QIGA) are found in various domains, including flow shop scheduling [89][90], power system optimization [91], thermal unit commitment [92], network design problems [93], etc. The NP-hard combinatorial optimization problem with strong engineering backgrounds [94], multiobjective flow shop scheduling problem (FSSP) [94] solved by the hybrid quantum-inspired genetic algorithm (HQGA) [94]. Han and Kim [95] proved that the quantum-inspired evolutionary algorithm (QEA) performed better than the conventional genetic algorithm on combinatorial optimization of the knapsack problem. The termination criterion, a two-phase scheme and Q-gate are changed to improve the presentation of

QEA. Konar *et al.* [96] proposed a Hybrid Quantum-Inspired Genetic Algorithm (HQIGA) in multiprocessor environment for real-time scheduling efficiently. Rotation gate used in HQIGA to investigate the variable chromosomes portrayed by qubits in Hilbert hyperspace. In this approach, random key distribution is utilized to change over the qubits chromosomes to legitimate timetable arrangements.

Economic load dispatch (ELD), a significant research field in the operation of thermal power plants, is a non-linear and complex optimization problem. Using ELD, power generation can be assigned to coordinate the load demand at negligible conceivable expense while fulfilling all the units and system constraints [97]. Due to different types of non-smooth cost function with equality and inequality constraints applied in the ELD problem, finding the global optimal solutions is very much difficult using the traditional approaches. Chaos Quantum-Inspired Particle Swarm Optimization (CQPSO) presented in [97] for ELD problem with better search and higher convergence ability. The concept of quantum computing as well as the implementations of self-adaptive probability selection and chaotic sequences mutation in this approach. Hassan *et al.* [98] presented quantum-inspired bat algorithm for solving economic load dispatch problem. Many-objective environmental economic dispatch (EED) issues can be unraveled by the quantum-inspired particle swarm optimization (QPSO) technique [99]. A definitive study of research works in solving fathoming different aspects of Economic Load Dispatch (ELD) problems of power system engineering using various kinds of PSO algorithms narrated in [100]. This article focused on five main areas of ELD problems [100], such as single objective economic load dispatch, economic load dispatch with non-conventional sources, dynamic economic load dispatch, multi-objective environmental/economic dispatch, and economic load dispatch of microgrids.

Wireless Sensor Network (WSN) [101], an emerging and fast growing technology in the context of Industry 4.0 [102], has been viewed as one of the predicting technologies in smart manufacturing. In this modern life, WSNs are applied to monitor the running equipment in a complex manufacturing environment. WSNs are utilized not exclusively to give administrations lifetime yet in addition to achieve fast and high-quality transmission of equipment monitoring data to monitoring centers [103]. The Quantum Ant Colony Multi-Objective Routing (QACMOR), a WSN routing algorithm, presented in mix with quantum computing and multi-objective fitness function into the routing research algorithm. It is developed to monitor the manufacturing environments. In this method, the node pheromone is represented by the quantum bits and the pheromone of the search path are updated by rotating the quantum gates. Ullah and Wahid [104] proposed a quantum-inspired genetic algorithm for topology control in the wireless sensor network. In this method, a linked quantum register that are binary pair of genes has been presented and its local proximity comprise the highly connected nodes and the energy consumption is very low. A quantum-inspired genetic algorithm can be proficient enough for grouping of WSNs to boost the lifetime of WSNs [105].

1.7 CONCLUSION

This chapter presents an outline of the basic theory and concept pertaining to quantum inspired metaheuristics. This chapter throws light on several types of quantum inspired metaheuristics in details. This chapter also comes up with a bird's eye view on different bi-level/multi-level quantum system-based optimization techniques. In addition to that, several entanglement induced optimization techniques and W-state encoding of optimization methods have also been discussed. The applications related to the theme of the topic have been provided that would also certainly bring up to date the readers.

REFERENCES

1. C. Blum and A. Roli. Metaheuristic in combinatorial optimization: Overview and conceptual comparison. Technical report, IRIDIA: Technical Report, 13, 2001.
2. F. Glover and G. A. Kochenberger. Handbook on Metaheuristics. Kluwer Academic Publishers, 2003.
3. K. Kennedy and R. Eberhart. Particle swarm optimization. In: Proceedings of the IEEE International Conference on Neural Networks (ICNN95), Perth, Australia, 4:1942–1948, 1995.
4. M. Dorigo, V. Maniezzo, and A. Colorni. The ant system: Optimization by a colony of cooperating agents. IEEE Transactions on Systems, Man, and Cybernetics, Part B, 26(1):29–41, 1996.
5. S. Kirkpatrik, C.D. Gelatt, and M.P. Vecchi. Optimization by simulated annealing. Science, 220:671–680, 1995.
6. F. Glover. Tabu search – Part I. ORSA Journal on Computing, 1(3):190–206, 1998.
7. F. Glover. Tabu search – Part II. ORSA Journal on Computing, 2(1):4–32, 1990.
8. F. Glover. Tabu search and adaptive memory programming: Advances, applications and challenges. In: Barr, Helgason, and Kennington, editors, Interfaces in Computer Science and Operations Research. Kluwer Academic Publishers, 1996.
9. R. Storn and K. Price. Differential evolution – a simple and efficient heuristic for global optimization over continuous spaces. Technical report, Technical Report TR-95-012, ICSI.
10. J. L. Cohon. Multiobjective Programming and Planning. Academic Press, New York, 1978.
11. D. E. Goldberg. Genetic Algorithms in Search, Optimization and Machine Learning. Addison-Wesley Longman Publishing Co., Inc., Boston, MA, 1989.
12. D. A. V. Veldhuizen and G. B. Lamont. Multiobjective evolutionary algorithms: Analyzing the state-of-the-art. Journal of Evolutionary Computation, 8(2):125–147, 2000.
13. K. Deb. Multi-objective optimization using evolutionary algorithms. Wiley, Chichester, UK, 1989.
14. L. K. Grover. Quantum computers can search rapidly by using almost any transformation. Physical Review Letters, 80(19):4329–4332, 1998.
15. P. W. Shor. Polynomial-time algorithms for prime factorization and discrete logarithms on a quantum computer. SIAM Journal of Computing, 26(5):1484–1509, 1997.
16. N. Metropolis, A. W. Rosenbluth, M.N. Rosenbluth, A. H. Teller, and E. Teller. Equation of state calculations by fast computing machines. The Journal of Chemical Physics, 21(6):1087–1092, 1953.
17. H. Lourenco, O. Martin, and T. St utzle. Iterated local search. Handbook on Metaheuristics, 2003.

18. C. Voudouris and E. Tsang. Guided local search and its application to the traveling salesman problem. European Journal of Operational Research, 113(2):469–499, 1999.
19. M. Dorigo, V. Maniezzo, and A. Colorni. The ant system: An autocatalytic optimizing process. Technical Report Technical Report TR91-016, Politecnico di Milano, Italy, 1991.
20. A. Colorni, M. Dorigo, and V. Maniezzo. Distributed optimization by ant colonies. In: Proceedings of ECAL91 European Conference on Artificial Life, pages 131–142, Elsevier, Amsterdam, The Netherlands, 1991.
21. L. Wang, Q. Niu, and M. R Fei. A novel quantum ant colony optimization algorithm and its application to fault diagnosis. Transactions of the Institute of Measurement and Control, 30(3–4):313–329, 2008.
22. S. Dey, S. Bhattacharyya, and U. Maulik. Quantum inspired meta-heuristic algorithms for multi-level thresholding for true colour images. In: Proceeding of 2013 Annual IEEE India Conference (INDICON), Mumbai, India, 2013.
23. S. Dey, S. Bhattacharyya, and U. Maulik. Efficient quantum inspired metaheuristics for multi-level true colour image thresholding. Applied Soft Computing, 56:472–513, 2017.
24. S. Dey, S. Bhattacharyya, and U. Maulik. New quantum inspired meta-heuristic techniques for multi-level colour image thresholding. Applied Soft Computing, 46:677–702, 2016.
25. T. A. Feo and M. G. C. Resende. Greedy randomized adaptive search procedures. Journal of Global Optimization, 6(2):109–133, 1995.
26. J. P. Hart and A. W. Shogan. Semi-greedy heuristics: An empirical study. Operations Research Letters, 6(3):107–114, 1987.
27. M. Prais and C. C. Ribeiro. Reactive grasp: An application to a matrix decomposition problem in TDMA traffic assignment. INFORMS Journal on Computing, 12(3):164–176, 2000.
28. M. G. C. Resende and C. C. Ribeiro. Greedy Randomized Adaptive Search Procedures. Handbook on Metaheuristics, Springer, 2003.
29. J. Holland. Adaptation in neural artificial systems. University of Michigan, Ann Arbor, MI, 1975.
30. H.-G. Beyer and H.-P. Schwefel. Evolution strategies: A comprehensive introduction. Journal Natural Computing, 1(1):3–52, 2002.
31. D. B. Fogel. Evolutionary Computation. IEEE Press, Piscataway, NJ, 1995.
32. F. Glover, M. Laguna, and R. Marti. Fundamentals of scatter search and path relinking. Control and Cybernetics, 39(3):653–684, 2000.
33. F. Glover, M. Laguna, and R. Marti. Scatter search and path relinking: Advances and applications. Handbook of metaheuristics, 2003.
34. I. Salman, O. Ucan, O. Bayat, and K. Shaker. Impact of metaheuristic iteration on artificial neural network structure in medical data. Processes, 6(57), 2018.
35. P. Bertolazzi, G. Felici, P. Festa, G. Fiscon, and E. Weitschek. Integer programming models for feature selection: New extensions and a randomized solution algorithm. European Journal of Operational Research, 250:389–399, 2016.
36. P. Festa and M. G. C. Resende. Basic components and enhancements. Telecommunication Systems, 46:253–271, 2011.
37. C. Blum, J. Puchinger, G.R. Raidl, and A. Roli. Hybrid metaheuristics in combinatorial optimization: A survey. Applied Soft Computing, 11:4135–4151, 2011.
38. G. Souza, E. Goldbarg, M. Goldbarg, and A. Canuto. A multiagent approach for metaheuristics hybridization applied to the traveling salesman problem. In: Proceedings of the 2012 Brazilian Symposium on Neural Networks, Curitiba, Parana, Brazil, 20–25 October 2012, pages 208–213, 2012.

39. Z. Tiejun, T. Yihong, and X. Lining. A multi-agent approach for solving traveling salesman problem. Wuhan University Journal of Natural Sciences, 11:1104–1108, 2006.

40. X.-F. Xie and J. Liu. Multiagent optimization system for solving the traveling salesman problem (TSP). IEEE Transactions on Systems, Man, and Cybernetics, Part B, 39:489–502, 2009.

41. F. Fernandes, S. Souza, M. Silva, H. Borges, and F. Ribeiro. A multiagent architecture for solving combinatorial optimization problems through metaheuristics. In: Proceedings of the IEEE International Conference on Systems, Man and Cybernetics, San Antonio, TX, USA, 11–14 October 2009, pages 3071–3076, 2009.

42. R. Malek. An agent-based hyper-heuristic approach to combinatorial optimization problems. In: Proceedings of the IEEE International Conference on Intelligent Computing and Intelligent Systems (ICIS), Xiamen, China, 29–31 October 2010, pages 428–434, 2010.

43. M. Milano and A. Role. MAGMA: A multiagent architecture for metaheuristics. IEEE Transactions on Systems, Man, and Cybernetics, Part B, 34:925–941, 2004.

44. G. X. Zhang. Quantum-inspired evolutionary algorithms: A survey and empirical study. Journal of Heuristics, 17(3):303–351, 2011.

45. G. X. Zhang. Time-frequency atom decomposition with quantum-inspired evolutionary algorithms. Circuits, Systems, and Signal Processing, 29(2):209–233, 2010.

46. M. D. Platel, S. Schliebs, and N. Kasabov. Quantum-inspired evolutionary algorithm: A multimodel EDA. IEEE Transactions on Evolutionary Computation, 13(6):1218–1232, 2009.

47. K. H. Han and J.-H. Kim. Quantum-inspired evolutionary algorithm for a class of combinatorial optimization. IEEE Transactions on Evolutionary Computation, 6(6):580–593, 2002.

48. K. H. Han and J.-H. Kim. Quantum-inspired evolutionary algorithms with a new termination criterion, h gate, and two-phase scheme. IEEE Transactions on Evolutionary Computation, 8(2):156–169, 2004.

49. S. Y. Kuo and Y. H. Chou. Entanglement-enhanced quantum-inspired Tabu search algorithm for function optimization. IEEE Access, 5, 2017.

50. H.-P. Chiang, Y.-H. Chou, C.-H. Chiu, S.-Y. Kuo, and Y.- M. Huang. A quantum-inspired Tabu search algorithm for solving combinatorial optimization problems. Soft Computing, 18(9):1771–1781, 2014.

51. Y.-H. Chou, S.-Y. Kuo, C.-Y. Chen, and H.-C. Chao. A rule-based dynamic decision-making stock trading system based on quantum-inspired Tabu search algorithm. IEEE Access, 2:883–896, 2014.

52. Y.-H. Chou, S.-Y. Kuo, C. Kuo, and Y.-C. Tsai. Intelligent stock trading system based on QTS algorithm in Japan's stock market. In: Proceedings of the IEEE International Conference on Systems, Man and Cybernetics, San Antonio, TX, pages 997–982, 2013.

53. W. Dur, G. Vidal, and J. I. Cirac. Three qubits can be entangled in two inequivalent ways. Physical Review A, 62(6), 2000.

54. E. DHondt and P. Panangaden. The computational power of the W and GHZ states. Quantum Information & Computation, 6(2), 2005.

55. X. B. Chen, Q. Y. Wen, F. Z. Guo, Y. Sun, G. Xu, and F. C. Zhu. Controlled quantum secure direct communication with W state. International Journal of Quantum Information, 6(4):899–906, 2008.

56. M. M. Cunha, A. Fonseca, and E. O. Silva. Tripartite entanglement: Foundations and applications. Universe, 5(209), 2019.

57. D. Cruz, R. Fournier, F. Gremion, A. Jeannerot, K Komagata, T. Tosic, J. Thiesbrummel, C. L. Chan, N. Macris, M. A. Dupertuis, and J. G. Clement. Efficient quantum algorithms

for GHZ and W states, and implementation on the IBM quantum computer. Advanced Quantum Technologies, 2(5–6), 2019.

58. S. A. MirHassani, S. Raeisi, and A. Rahmani. Quantum binary particle swarm optimization-based algorithm for solving a class of bi-level competitive facility location problems. Optimization Methods and Software, 30(4):756–768, 2015.

59. G. Zhang, G. Zhang, Y. Gao, and J. Lu. Competitive strategic bidding optimization in electricity markets using bi-level programming and swarm technique. IEEE Transactions on Industrial Electronics, 58(6):2138–2146, 2011.

60. S. Dey, S. Bhattacharyya, and U. Maulik. Quantum inspired genetic algorithm and particle swarm optimization using chaotic map model based interference for gray level image thresholding. Swarm and Evolutionary Computation, 15:38–57, 2014.

61. T. Zhang, T. Hu, J. W. Chen, Z. Wan, and X. Guo. Solving bi-level multiobjective programming problem by elite quantum behaved particle swarm optimization. Abstract and Applied Analysis, 2012, 2012.

62. S. Kumar, P. Kumar, T. K. Sharma, and M. Pant. Bi-level thresholding using PSO, artificial bee colony and MRLDE embedded with Otsu method. Memetic Computing, 5:323–334, 2013.

63. X. Chang, Z. Ma, Y. Yang, Z. Zeng, and A. G. Hauptmann. Bi-level semantic representation analysis for multimedia event detection. Memetic Computing, 27(5):1180–1197, 2017.

64. X. Yan, N. Lv, Z. Liu, and K. Xu. Quantum-inspired evolutionary algorithm for transportation network design optimization. In: Proceeding of 2008 Second International Conference on Genetic and Evolutionary Computing, Hubei, China, 2008.

65. S. Dey, S. Bhattacharyya, and U. Maulik. Quantum-inspired multi-objective simulated annealing for bi-level image thresholding. Quantum Inspired Computational Intelligence, Research and Applications, 2017.

66. S. Dey, I. Saha, U. Maulik, and S. Bhattacharyya. New quantum inspired meta-heuristic methods for multi-level thresholding. In: Proceeding of 2013 International Conference on Advances in Computing, Communications and Informatics (ICACCI), Mysore, India, 2013.

67. V. Tkachuk. Quantum genetic algorithm on multilevel quantum systems. Mathematical Problems in Engineering, 2018:1–12, 2018.

68. F. A. Cardenas-Lopez, L. Lamata, J. C. Retamal, and E. Solano. Multiqubit and multilevel quantum reinforcement learning with quantum technologies. PLoS ONE, 13(7):1–12, 2018.

69. P. Niemann, R. Wille, and R. Drechsler. Equivalence checking in multi-level quantum systems. International Conference on Reversible Computation, LNCS, 8507:201–215, 2014.

70. S. Carrasco, J. Rogan, and J. A. Valdivia. Speeding up maximum population transfer in periodically driven multi-level quantum systems. Scientific Reports, 9, 2019.

71. M. Grace, C. Brif, H. Rabitz, I. Walmsley, R. Kosut, and D. Lidar. Encoding a qubit into multilevel subspaces. New Journal of Physics, 8, 2006.

72. B. C. Roy and P. K. Das. Optimal control of multi-level quantum system with energy cost functional. International Journal of Control, 80(8):1299–1306, 2007.

73. J. Cao H. Gao and M. Diao. A simple quantum-inspired particle swarm optimization and its application. Information Technology Journal, 10(12):2315–2321, 2011.

74. H.-P. Chiang, Y.-H. Chou, C.-H. Chiu, S.-Y. Kuo, and Y.-M. Huang. A quantum-inspired Tabu search algorithm for solving combinatorial optimization problems. Soft Computing, 18(9):1771–1781, September 2014.

75. S. Kuo and Y. Chou. Entanglement-enhanced quantum-inspired Tabu search algorithm for function optimization. IEEE Access, 5:13236–13252, 2017.

76. H. Wang, J. Liu, J. Zhi, and C. Fu. The improvement of quantum genetic algorithm and its application on function optimization. Mathematical Problems in Engineering, 2013(730749):1–10, 2013.

77. H. Xiong, Z. Wu, H. Fan, G. Li, and G. Jiang. Quantum rotation gate in quantum-inspired evolutionary algorithm: A review, analysis and comparison study. Swarm and Evolutionary Computation, 42:43–57, 2018.

78. D. Zouache, F. Nouioua, and A. Moussaoui. Quantum-inspired firefly algorithm with particle swarm optimization for discrete optimization problems. Soft Computing, 20(7):2781–2799, July 2016.

79. H. N. Abdull Hamed, N. Kasabov, Z. Michlovsky, and S. M. Shamsuddin. String pattern recognition using evolving spiking neural networks and quantum inspired particle swarm optimization. In: Neural Information Processing. Lecture Notes in Computer Science, vol. 5864, pages 611–619. Springer, Berlin, Heidelberg, 2009.

80. J. Zhang, H. Li, Z. Tang, Q. Lu, X. Zheng, and J. Zhou. An improved quantum-inspired genetic algorithm for image multilevel thresholding segmentation. Mathematical Problems in Engineering, 2014(295402):1–12, 2014.

81. S. Dey, S. De, D. Ghosh, D. Konar, S. Bhattacharyya, and J. Platos. A novel quantum inspired sperm whale metaheuristic for image thresholding. In: 2019 Second International Conference on Advanced Computational and Communication Paradigms (ICACCP), pages 1–7, Feb 2019.

82. S. Das, S. De, and S. Bhattacharyya. True color image segmentation using quantum-induced modified-genetic-algorithm-based FCM algorithm. In: Quantum-Inspired Intelligent Systems for Multimedia Data Analysis, pages 55–94. Research Essentials Collection. IGI Global, 2018.

83. S. Das, S. De, S. Bhattacharyya, and A. E. Hassanien. Color MRI image segmentation using quantum-inspired modified genetic algorithm-based FCM. In: Recent Trends in Signal and Image Processing. Advances in Intelligent Systems and Computing, vol. 727, pages 151–164. Springer, Singapore, 2019.

84. F. Liu, H. Duan, and Y. Deng. A chaotic quantum-behaved particle swarm optimization based on lateral inhibition for image matching. Optik, 123(12):1955–1960, 2012.

85. Y. Feng, H. Yin, H. Lu, L. Cao, and J. Bai. FCM-based quantum artificial bee colony algorithm for image segmentation. In: Proceedings of the 10th International Conference on Internet Multimedia Computing and Service, pages 6:1–6:7, 2018.

86. E. Osaba, J. Del Ser, A. Iglesias, and X.-S. Yang. Soft computing for swarm robotics: New trends and applications. Journal of Computational Science, 39, 2020.

87. Y. Tan and Z. Zheng. Research advance in swarm robotics. Defence Technology 9:18–39, 2013.

88. M. Contreras-Cruz, V. Ayala, and U. Hernandez-Belmonte. Mobile robot path planning using artificial bee colony and evolutionary programming. Applied Soft Computing, 30:319–328, 2015.

89. J. Gu, X. Gu, and B. Jiao. A quantum genetic based scheduling algorithm for stochastic flow shop scheduling problem with random breakdown. IFAC Proceedings Volumes, 41(2):63–68, 2008.

90. J. Gu, X. Gu, and M. Gub. A novel parallel quantum genetic algorithm for stochastic job shop scheduling. Journal of Mathematical Analysis and Applications, 355(1):63–81, 2009.

91. A. K. Al-Othman, F. S. Al-Fares, and K. M. EL-Nagger. Power system security constrained economic dispatch using real coded quantum inspired evolution algorithm. International Journal of Nuclear and Quantum Engineering, 1(5):4–10, 2007.

92. Y.-W. Jeong, J.-B. Park, J.-R. Shin, and K. Y. Lee. A thermal unit commitment approach using an improved quantum evolutionary algorithm. Electric Power Components and Systems, 37(7):770–786, 2009.

93. H. Xing, X. Liu, X. Jin, L. Bai, and Y. Ji. A multi-granularity evolution based quantum genetic algorithm for QoS multicast routing problem in WDM networks. Computer Communications, 32(2):386–393, 2009.

94. B. Li and L. Wang. A hybrid quantum-inspired genetic algorithm for multiobjective flow shop scheduling. IEEE Transactions on Systems, Man, and Cybernetics, Part B (Cybernetics), 37(3):576–591, June 2007.

95. K.-H. Han and J.-H. Kim. Quantum-inspired evolutionary algorithms with a new termination criterion, h/sub /spl epsi// gate, and two-phase scheme. IEEE Transactions on Evolutionary Computation, 8(2):156–169, April 2004.

96. D. Konar, S. Bhattacharyya, K. Sharma, S. Sharma, and S. R. Pradhan. An improved hybrid quantum-inspired genetic algorithm (HQIGA) for scheduling of real-time task in multiprocessor system. Applied Soft Computing, 53:296–307, 2017.

97. K. Meng, H. G. Wang, Z. Dong, and K. P. Wong. Quantum-inspired particle swarm optimization for valve-point economic load dispatch. IEEE Transactions on Power Systems, 25(1):215–222, Feb 2010.

98. H. Tehzeeb ul Hassan, M. U. Asghar, M. Z. Zamir, and H. M. A. Faiz. Economic load dispatch using novel bat algorithm with quantum and mechanical behaviour. In: 2017 International Symposium on Wireless Systems and Networks (ISWSN), pages 1–6, Nov 2017.

99. F. Mahdi, P. Vasant, M. Abdullah-Al-Wadud, J. Watada, and V. Kallimani. A quantum-inspired particle swarm optimization approach for environmental/economic power dispatch problem using cubic criterion function. International Transactions on Electrical Energy Systems, page e2497, 12 2017.

100. M. N. Alam. State-of-the-art economic load dispatch of power systems using particle swarm optimization. CoRR, abs/1812.11610, 2018.

101. C. Li, X. Xie, Y. Huang, H. Wang, and C. Niu. Distributed data mining based on deep neural network for wireless sensor network. International Journal of Distributed Sensor Networks, 11(7):1–7, 2015.

102. Xiaomin Li, D. Li, J. Wan, A. V. Vasilakos, C.-F. Lai, and S. Wang. A review of industrial wireless networks in the context of industry 4.0. Wireless Networks, 23(1):23–41, 2017.

103. F. Li, M. Liu, and G. Xu. A quantum ant colony multi-objective routing algorithm in WSN and its application in a manufacturing environment. Sensors, 19(15), 2019.

104. S. Ullah and M. Wahid. Topology control of wireless sensor network using quantum inspired genetic algorithm. International Journal of Swarm Intelligence and Evolutionary Computation, 04, 08 2015.

105. M. Rathee and S. Kumar. Quantum inspired genetic algorithm for energy efficient clustering in wireless sensor networks. In: 2016 IEEE 1st International Conference on Power Electronics, Intelligent Control and Energy Systems (ICPEICES), pages 1–6, 07 2016.

2 A Quantum-Inspired Approach to Collective Combine Basic Classifiers in an Ensemble Bagging

2.1 INTRODUCTION

Machine learning algorithms are used to solve classification problems (supervised learning), clustering (unsupervised learning), reinforcement learning, etc. Learning consists in using the available data to adjust the variable parameters of the algorithm in order to fit the algorithm to the specifics of the problem being solved. Hyperparametric optimization (the problem of choosing a model) is the problem of finding a set of hyperparameters of a learning algorithm, at which the highest efficiency of the algorithm is achieved when solving a specific problem. The choice of a suitable method and its tuning have the greatest impact on the success of the solution of the task at hand. The performance of machine learning algorithms depends on the parameters of the algorithm and the chosen structure of the model. To solve the problem of optimal choice of a model, a number of methods have been proposed, which can be divided into 2 groups: hyperparametric optimization algorithms and ensemble methods, in which several machine learning algorithms are used in parallel to collectively solve the problem. Although none of the proposed algorithms is a panacea and does not guarantee a successful solution to the problem, experimental evidence of their effectiveness was obtained [1][2]. To collectively combine the decisions of the base classifiers, the voting method is usually used. In this chapter, we will look at alternative methods of combining such as stacking and ensemble selection, and also propose a new quantum-inspired approach based on metaheuristic hyperparameter tuning algorithms.

2.2 BAGGING METHOD

The Bootstrap aggregating method was proposed by Leo Breiman to improve classification efficiency by combining classifiers built on random subsamples of a given training sample.

Let there be a training sample X of size n. The bagging procedure generates L new training samples X_i, each of size n, by uniform selection with substitution from the original sample. Thus, the elements of the X_i samples may contain duplicates. The probability of being sampled (a subsample of size L from the main sample "with return") for sufficiently large samples is approximately 0.632, i.e. about 36.8% of

DOI: 10.1201/9781003283294-2

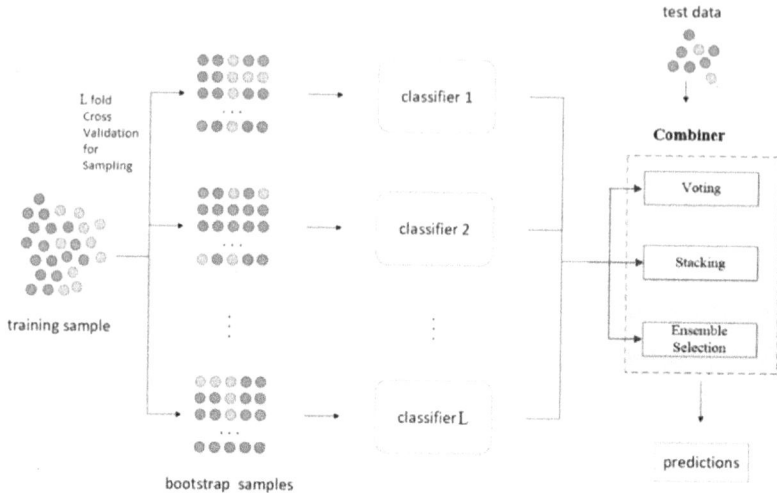

Figure 2.1: Bagging with final ensemble combiner (Adapted from Gorodetsky, V. & Serebryakov, S. (2006). Methods and algorithms of the collective recognition: a survey. SPIIRAS Proceedings, 3(1), 139-171)

examples from the training sample may be unclaimed. This type of selection is called Bootstrap. The L obtained samples are used to train L models, which in turn are combined into a team. In regression problems, the outputs of the classifiers are averaged; in classification problems, the voting method is usually used.

Bagging leads to improvements in unstable algorithms such as artificial neural networks and classification and regression trees (CARTs). Due to the use of bagging in some works, an improvement in pattern recognition was noted [3][4]. Thus, bagging is useful in the case of different classifiers and instability, when small changes in the initial sample lead to significant changes in the classification [5].

As seen in Figure 2.1, the final ensemble combiner can use different ways to combine the classifier decisions, including voting, stacking and ensemble selection. Existing classification algorithms can be combined into several groups according to the principle of their operation:

1. Classifiers based on the similarity of objects.
2. Algorithms for statistical classification.
3. Classifiers based on the separability of classes in the attribute space.
4. Logical classification algorithms.
5. Neural networks.

2.3 CLASSIFIERS BASED ON SIMILARITY OF OBJECTS

The algorithms of this group are based on the compactness hypothesis, which assumes that objects of the same class are most often similar to each other, and objects of different classes are different. The similarity of objects in an n-dimensional feature space can be calculated using various metrics:

1. Euclidean distance:

$$d_E(x,y) = \sqrt{\sum_i^n (x_i - y_i)^2} \qquad (2.1)$$

2. Manhattan distance:

$$d_{Man}(x,y) = \sum_i^n |x_i - y_i|^2 \qquad (2.2)$$

3. Chebyshev distance:

$$d_{Ch}(x,y) = max_i(|x_i - y_i|) \qquad (2.3)$$

4. Minkowski distance of the pth order:

$$d_{Mink}(x,y) = (\sum_i^n |x_i - y_i|^p)^{\frac{1}{p}} \qquad (2.4)$$

The most famous algorithms in this group are the following:

1. The method of constructing standards: Based on the training sample, reference points are calculated, which are the centers of objects of each class in the attribute space, according to the formula:

$$E_i^k = \frac{1}{n^k} \sum_i (x_i : f(\bar{x}) = k); i = 1, 2, \ldots m; k = 1, 2, \ldots K \qquad (2.5)$$

where, E_k is the standard of the kth class; n_k is the number of objects in the sample of the kth class; m is the dimension of the feature space and K is the number of classes.
According to a certain metric, the distance from the classified object to each of the standards is calculated, the object is attributed to the class, the distance to the standard of which is minimal.

2. Method of k nearest neighbors [6]: For this algorithm, the data must be represented as a matrix of distances between the sampled objects, calculated according to a certain metric. Classes of k objects closest to the classified object are considered. The object belongs to the class that is most often found among its neighbors. When solving practical problems with this method, it is very important to choose the correct metric for calculating the distance between objects, as well as the value of the parameter k. If the value of the parameter is too small (for example, $k = 1$), the algorithm becomes susceptible to the negative influence of outliers; if the value of k is too high, too many neighboring objects are included in the calculations, which can also negatively affect the quality of the classification, since among the neighbors of the classified object there may be many objects of another class.

3. The method of potential functions [7]: This method is based on the physical principle of the potential of the electric field of a charged particle. The distance from the classified object to each object of the training sample is calculated. The decision rule is constructed as in the method of nearest neighbors, the difference is that the sample object has some measure of importance ("charge") relative to the classified object.

2.4   STATISTICAL CLASSIFICATION ALGORITHMS

Statistical classification algorithms are based on estimating the distribution density of classes over a sample. Depending on the estimation method, algorithms are distinguished with parametric and nonparametric density estimation, as well as density estimation as a mixture of parametric distributions. Well-known algorithms of this group:

1. Naive Bayesian classifier [8]: This method is based on the assumption that the features that describe the objects of the sample are statistically independent. This assumption greatly facilitates the problem of estimating the distribution density, since instead of the n-dimensional density, it is necessary to estimate n one-dimensional densities. Densities can be estimated both parametrically and nonparametrically. According to the Bayes rule, the posterior probabilities of each of the K classes are found, provided that the attribute x of the classified object is measured:

$$P(i|x) = \frac{f(x|i) \times P(i)}{\sum_j^m f(x|j) \times P(j)}; i = 1, 2, \ldots K \qquad (2.6)$$

where, $f(x|i)$ is the assessment of the conditional distribution density of the attribute x for the ith class and $P(i)$ is an estimate of the prior probability of the class.

The decision rule for the classified object x is as follows:

$$i* = arg\ max_i P(i|x); i = 1, 2, \ldots K \qquad (2.7)$$

2. Parzen window method: This method uses nonparametric estimation of the density [9] of the distribution of classes for the available sample, therefore, it does not put forward hypotheses about the structure of the distribution density function. The decision rule for classifying object x is as follows:

$$i* = arg\ max_i \lambda_i \sum_{j=1}^n [y_j = i] K(\frac{d(x, x_j)}{h}); i = 1, 2, \ldots K \qquad (2.8)$$

where, λ_i is the price of the correct answer for class i; n is the sample size; y_j is the class of the jth object; $K(t)$ is the nuclear function; $d(x, x_j)$ is the distance between the classified object x and the object x_j and h is the window width.

3. EM-algorithm (expectation-maximization): The EM algorithm [10] estimates the density as a mixture of parametric distributions. In this algorithm, two stages are

iteratively performed: the estimation stage, in which the expected value of the likelihood function is calculated, and the maximization stage, in which the parameters of the likelihood function are calculated that maximize it.

2.5 CLASSIFIERS BASED ON CLASS SEPARABILITY IN ATTRIBUTE SPACE

These algorithms build a dividing surface in the feature space, dividing objects into disjoint classes. The most famous methods of this group are as follows:

1. Fisher's linear discriminant [11], also known as linear discriminant analysis, is applicable if the sample satisfies the following hypotheses: the classes are normally distributed and the class covariance matrices are equal. Fisher's linear discriminant is a simplification of the quadratic discriminant. In the case of two classes in two-dimensional space, the dividing surface constructed using this method will be a straight line. In the case of a larger number of classes, the dividing surface will be piecewise linear.

2. Logistic regression [12]. For the case of 2 classes, a linear classification algorithm is constructed with a decision rule of the form:

$$log\ reg(x,w) = sign(\sum_{j=1}^{m} w_j x_j - w_0) = sign(x,w) \qquad (2.9)$$

where, w_j is the weight of the jth feature; w_0 is the decision threshold; w is the vector of weights and (x,w) is the scalar multiplication of the weights vector and features of the object.

The problem of training the logistic regression algorithm is to find the optimal vector of weights w that minimizes the loss function of the form:

$$L(w) = \sum_{i=1}^{n} ln(1 + \frac{1}{e^{y_i(x,w)}}) \qquad (2.10)$$

3. The support vector machine [13] is one of the most popular supervised learning methods for several reasons:
(i) The fastest method for finding the decision rule.
(ii) Reduces to solving a quadratic programming problem in a convex domain, which always has a unique solution.
(iii) Finds the dividing surface of the classes with a dividing bar of maximum width, which contributes to more confident classification.
In the case of two classes and a linearly separable sample, the decision rule of the SVM algorithm takes the form:

$$y_i((w,x_i) + b) \geq 1; i = 1, 2, ... n \qquad (2.11)$$

where, y_i is the class label of the ith sample object; w is the vector of weight coefficients and n is the sample size.

The vector w is sought in the form:

$$w = \sum_{i=1}^{n} \lambda_i y_i x_i \tag{2.12}$$

Into this sum with nonzero coefficients λ_i includes only those objects of the selection that lie on the dividing surface. These objects are called support vectors. In the case of a linearly inseparable sample, the feature space R^n is transferred to a space of higher dimension H using function $\phi(x)$, in which the sample becomes linearly divisible. In this case, the decision rule is sought in the form:

$$y_i((w, \phi(x_i)) + b) \geq 1; i = 1, 2, \ldots n \tag{2.13}$$

$$w = \sum_{i=1}^{n} \lambda_i y_i \phi(x_i) \tag{2.14}$$

There is also a multiclass support vector machine, which is reduced to dividing the problem into several binary classification problems according to the "one against all" scheme.

2.6   LOGICAL CLASSIFICATION ALGORITHMS

The principle of operation of logical classification algorithms is based on the construction of hierarchical compositions of simple rules. The most prominent representative of this group of algorithms is a decision tree [14]. The structure of the tree consists of the so-called branches and leaves, in the branches signs are written on which the objective function depends, and in the leaves - the values of the objective function. Decision trees are suitable for both classification problems and regression problems. In the case of the classification problem, the class to which the classification object belongs is recorded in the leaves of the tree. In the case of regression problems - a real number, The tree construction scheme looks like this:

1. The next feature fi(x) is selected.
2. The values of the selected feature are divided into several subgroups, thereby dividing the feature space into several subspaces.
3. The procedure is recursively repeated until one of the stopping criteria is reached - the maximum tree depth, the number of leaves, or the maximum size of each leaf.

Depending on the principle by which the next feature for splitting is selected, and how it is split, there are several variants of this algorithm:

1. Algorithm ID3 - in it, the next feature is selected according to the criterion of information gain.
2. Algorithm C4.5 [15] is an improved version of ID3, the choice of a feature according to the criterion of normalized increment of information.

3. Algorithm CART (classification and regression tree) [16].

In practice, in order to avoid the effect of overfitting after building a decision tree, some of its branches are truncated to maintain better generalizing ability, this procedure is called pruning.

2.7 NEURAL NETWORKS

A group of classification algorithms called neural networks [17] is based on the principle of linking a large number of simple elements built in the form of an optimal structure. Artificial neurons, mathematical constructs inspired by the study of nerve cells in the human brain, act as simple elements. The concept of a neural network was formalized by McCulloch and Pitts in 1943, and the first learning algorithm was proposed by Hebb in 1949. In 1958, Rosenblatt invented the perceptron, the first neural network used for pattern recognition. In those years, there was a high interest in neural networks in scientific circles, but after Minsky in 1969 published a formal proof of the limitations of the perceptron in recognizing complex, variable images, interest in neural networks in general, and to the perceptron in particular, it dropped sharply. A little later, in 1974, an error backpropagation algorithm was invented, and in 1982 neural networks with feedback were proposed, the training of which was carried out on the basis of minimizing the so-called network energy function. In 1986, interest in the direction of neural networks was renewed after the backpropagation algorithm was significantly improved. Later, this algorithm was successfully applied to train a new generation of deep neural networks.

2.8 METHODS OF COMBINING BASIC CLASSIFIERS

As seen in 9.9, the final ensemble combiner can use different ways to combine the classifier decisions, including voting, stacking and ensemble selection. Those combination methods are briefly introduced as follows.

2.8.1 VOTING

Let a classifier ensemble consists of L base classifiers in the set $D = \{D_1, D_2, \ldots, D_L\}$, and any object $x \in R^n$ is assigned to one of the c possible classes $\Omega = \{\omega_1, \omega_2, \ldots, \omega_c\}$.

For x to be classified, L classifiers output a matrix $M = [m_{i,j}]; i = 1, 2, \ldots, L; j = 1, 2, \ldots, c$.

1. Majority voting rule: Suppose $m_{i,j} \in \{0, 1\}$, where $m_{i,j} = 1$ if D_i predicts x in class ω_j and $m_{i,j} = 0$, otherwise. x is assigned to ω_k if

$$\sum_{i=1}^{L} m_{i,k} = \max_{j=1}^{c} \sum_{i=1}^{L} m_{i,j} \qquad (2.15)$$

2. Average of probabilities rule: Suppose $m_{i,j} \in \{0,1\}$, where $m_{i,j}$ is the degree of support that classifier D_i gives to the hypothesis that x comes from class ω_j, denoted as $m_{i,j} = P_{D_i}(\omega_i|x)$. x is assigned to ω_k if

$$\frac{1}{L}\sum_{i=1}^{L} m_{i,k} = \max_{j=1}^{c} \frac{1}{L}\sum_{i=1}^{L} m_{i,j} \qquad (2.16)$$

2.8.2 STACKING

Stacking constructs a set of heterogonous or homogeneous base classifiers, and the outputs of base classifiers are used to train metaclassifier which produces single output as final classification result. The task of the metaclassifier is to correct any mistakes made by base classifiers and minimize the generalization error. Any classification algorithm can be used to train base classifier or metaclassifier. The procedure of stacking algorithm is as follows:

1. Step 1: split a dataset into three disjoint subsets: the training set, the validation set, and the testing set;
2. Step 2: train a set of base classifiers on the training set;
3. Step 3: apply those base classifiers to classify the validation set;
4. Step 4: using the outputs of base classifiers from Step 3 as the features, along with the true class label, to train the metaclassifier;
5. Step 5: test the metaclassifier on the testing set to evaluate the performance of Stacking ensemble.

2.8.3 ENSEMBLE SELECTION

Most ensemble algorithms combine all base classifiers to construct an ensemble, but it is difficult to determine that what base classifier should be included in the ensemble, how many base classifiers are needed, and what ensemble strategy should be used to combine base classifiers.

Combining a portion of base classifiers instead of using all of them to construct an ensemble may be better. Such an ensemble is called as ensemble selection, which can achieve strong generalization performance with small size of base classifiers. As seen in Figure 2.2, the ensemble selection is built in three steps.

1. Step 1: A set of heterogonous or homogeneous base classifiers are trained for the same task.
2. Step 2: The chosen algorithm is employed to compute the weight of base classifiers, and a subset of base classifiers whose weight is bigger than a preset threshold is combined to construct an ensemble.
3. Step 3: The output layer calculates the final degree of membership Y of jth example x_j to the class base on Majority voting rule (soft vote) or Average of probabilities rule.

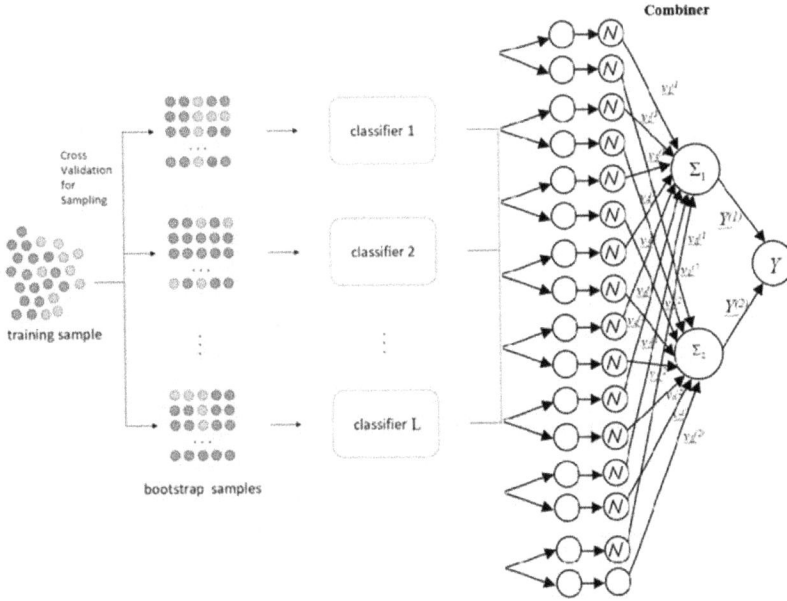

Figure 2.2: Ensemble selection (Modified from Gorbachev, S., Arkhipov, A., Gorbacheva, N., Bhattacharyya, S., Cao, J. & Kale, S. (2021). Study and Developing of Diversity Generation Methods in Heretogeneous Ensemble Models. International Journal of Distributed Computing and Technology, 7(1), 816)

The outputs $Y_j^{(k)}$ of the penultimate layer (Figure 2.2) for each ith class calculate the degree of membership of the jth example x_j of training sample to the class as a weighted average linear combination of the normalized outputs of each classifier (neuroexpert):

$$Y_j^{(1)} = \frac{1}{L}\sum_{i=1}^{L} v_i^{(1)} y_{ij}^{(1)}, \ldots Y_j^{(c)} = \frac{1}{L}\sum_{i=1}^{L} v_i^{(c)} y_{ij}^{(c)} \qquad (2.17)$$

The weights of each classifier (neuroexpert) are calculated based on the number of errors they made using the following methods:
(i) Fisher linear discriminant
(ii) logistic regression
(iii) Single-layer perceptron
(iv) SVM support vector machine, which is most significant in terms of maximizing the separation ability between classes and in terms of reliability
(v) the <<naive>> Bayesian classifier (the most popular of the simple ones)
(vi) heuristic algorithms.

2.8.3.1 The <<naive>> Bayesian Classifier

Let $y_1^{(1)}(x_j), \ldots y_L^{(1)}(x_j), \ldots y_1^{(c)}(x_j), \ldots y_L^{(C)}(x_j)$ be independent random variables. Then the weighted voting coefficients $v_i^{(k)}$ $(i = 1, 2, \ldots L)$ for each kth class $(k = 1, 2, \ldots c)$ can be calculated using the formula:

$$v_i = ln\frac{1 - p_i}{p_i}; i = 1, 2, \ldots L \tag{2.18}$$

where p_i is the probability of errors of the ith classifier on the training sample. The fewer errors it makes, the greater its weight. As a probability estimate, we can take the error rate γ_i or $\gamma_i + \frac{1}{N}$, so that the denominator does not vanish, i.e.:

$$v_i = ln\frac{1 - \gamma_i}{\gamma_i + \frac{1}{N}}; i = 1, 2, \ldots L \tag{2.19}$$

If any neuroexpert makes more than half of the mistakes, then his weight is $v_i < 0$. Such an unreliable classifier is not taken into account in the meta-network, putting $v_i = 0$.

2.8.4 QUANTUM-INSPIRED METAHEURISTICS METHOD

This section provides an overview of population-based quantum inspired metaheuristics. A group of researchers have intended to design a variety of quantum inspired algorithms when no quantum computers were available in the literature. There exist some appealing features in quantum computers that may provide remarkable outcomes when executing quantum inspired algorithms in classical computers. Nowadays, first generation of quantum computers are at hand, different scientists throughout the globe are working hard to assess the quantum efficacy in a variety of fields. A large variety of research works have already been introduced since the invention of first version of quantum behaved of metaheuristics. Metaheuristics are stochastic kind of algorithms, in which, based on some predefined policies, numerous heuristic algorithms are constructed [18]. Several problem-independent, efficient, popular approximate metaheuristics are available in the literature that can efficiently deal with different optimization problems. Metaheuristics are referred to as the high-level algorithmic structure that come up with a set of predefined strategies to introduce heuristic optimization techniques [19].

    An approximation technique is a process that provides approximate results to any specific problem. An approximation technique runs in polynomial time depending on the size of input variable, rather than exponential time to search for optimal solution (or near optimal solution). Execution time is a crucial metric for developing an approximation technique. This metric is highly influenced in positive direction with the advancement of computer architecture. Nowadays, algorithms are designed and modified in such a way that they can perform parallel processing. The multicore microprocessor is new generation hardware device which are frequently and efficiently used for executing these new versions of algorithms. Apart from that GPU cards which can accommodate plenty of processing units and CUDA (a specialized software) are also very useful for the same purpose [20]. In the recent times, a number of renowned companies have successfully designed powerful quantum computers which are more efficient than classical computers in all respects. These quantum computers can efficiently be used in different fields such as machine learning, simulations, optimization to name a few. Some of the popular companies which have successfully developed quantum computers are Google [21], IBM [22], Intel [23] and D-Wave Systems [24] to name a few. In the literature, some popular metaheuristics can be listed as simulated annealing (SA) [25], particle swarm optimization (PSO) [26], differential evolution (DE) [27], and ant colony optimization (ACO) [28] to name a few.

In 1996, Narayanan and Moore used the thought and features of quantum mechanics to develop efficient metaheuristics [22]. The authors used a quantum-inspired crossover to find optimal solution for the traveling salesman problem (TSP). A set of guidelines has been introduced as an effort to characterize a method for designing and developing quantum algorithms. The theory and feathers of quantum computing have been used by different researchers several times to design a number of quantum inspired algorithms. One pioneer algorithms of this category is popularly known as Genetic Quantum Algorithm (GQA) [30]. In 2019, Montiel *et al.* presented a popular quantum-inspired algorithm, called quantum-inspired Acromyrmex evolutionary algorithm (QIAEA) [31]. The authors carefully observed the colony habits of the Atta and Acromyrmex in their daily life. This fact motivated them to develop QIAEA.

Han *et al.* [32] developed a programmed version (parallel) of QGA, called parallel QGA (PQGA). PQGA has been applied on knapsack problem for optimization. PQGA has been compared with QGA to judge its efficacy, where PQGA outperformed other. In 2012, Li *et al.* [33] presented a watershed-based quantum evolutionary algorithm for texture image clustering and SAR segmentation. Wang *et al.* [34] introduced a QEA and particle swarm optimization method (PSO) based quantum swarm evolutionary algorithm (QSwE). A variety of Q-bit expression form and an improved version of PSO have been introduced in QSwE for updating quantum angle. Zouache *et al.* [35] introduced PSO based quantum-inspired firefly algorithm (QIFAPSO). The communal habits of the firefly, swarm and the concept of quantum computing have been combined in a single skeleton to design QIFAPSO.

Apart from that, several quantum based metaheuristics are available in the literature. Some of them are quantum-inspired evolutionary algorithm based on p-system (QEPS) [36] developed by Zhang *et al.*, quantum-inspired DE and PSO algorithm (QDEPSO) [37] proposed by Zouache and Moussaoui etc. Chaos Quantum-Inspired Particle Swarm Optimization (CQPSO) was developed in [38] to handle Economic load dispatch (ELD) problem. Hassan *et al.* [39] proposed quantum inspired bat algorithm to deal with various economic load dispatch problem. In the wireless sensor network, Ullah and Wahid [40] designed a quantum inspired genetic algorithm framework for topology control. Several quantum inspired algorithms have been introduced in the literature so far for handling bi-level optimization problem. Zhang *et al.* [41] used bi-level programming in collaboration with swarm intelligence to develop a strategic bidding optimization algorithm. Dey *et al.* [42] presented quantum inspired bi-level optimization algorithms using GA and PSO for gray-level image thresholding. The quantum inspired multi-objective based simulated annealing has been designed by Dey *et al.* [3] for bi-level image thresholding. The computational capability of bi-level system has been enhanced to the multi-level frame by altering its basic structure. Dey *et al.* [2] introduced Quantum inspired particle swarm optimization and quantum inspired differential evolution for multi-level colour image thresholding. Tkachuk [45] used quantum technological approach to develop quantum inspired evolutionary algorithm. Later, Dey *et al.* [2][4][47] designed a number of quantum inspired metaheuristics in multi-level and colour domain. Cardenas et al. [48] designed a protocol to perform quantum reinforcement learning (QRL) and quantum technologies (QT).

Quantum inspired metaheuristics have been widely used in pattern recognition. Dey *et al.* [49] introduced a quantum inspired sperm whale algorithm for multi-level thresholding. The basic operators of quantum computing have been fused with the sperm whale algorithm. Dutta *et al.* [50] introduced a novel metaheuristic, called the Border Collie Optimization. The motivation behind designing of this algorithm is mimicking the sheep herding behaviour of Border Collie dogs.

Unlike, single objective optimization (SOO), multi-objective optimization (MOO) handles more than one objective function at a time. Kim *et al.* [51] designed a QEA based quantum-inspired multi-objective evolutionary algorithm (QMEA) for solving the 0-1 knapsack problem. Moghadam *et al.* [52] first introduced a quantum version of gravitational search algorithm (GSA), called quantum-behaved gravitational search algorithm (QGSA).

Later, in 2015, Chakraborti *et al.* [53] introduced a modified version of QGSA, called modified binary quantum-behaved gravitational search algorithm with differential mutation (MBQGSA-DM). Like QGSA, this algorithm also used differential mutation strategy. Li and Wang [54] proposed a hybrid quantum-inspired genetic algorithm (HQGA). This algorithm has been designed to efficiently deal with multi-objective based combinatorial optimization problem, called low shop scheduling problem (FSSP). A novel algorithm, called Quantum Ant Colony Multi-Objective Routing (QACMOR) has been developed to deal with WSN routing problem. In this algorithm, the concepts of quantum computing and multi-objective function have been utilized.

2.9 CONCLUSION

With the development of machine learning theory and the accumulation of practical experience of using various algorithms, it became clear that there is no ideal classification method that would be better than all others for all sizes of the training sample, for any percentage of noise in data, for any complexity of the boundaries of dividing objects into classes etc. Therefore, at present, ensemble classification methods that combine many different classifiers trained on different data samples. One of the most accurate and fast parallelization methods available today is bagging, which turns out to be useful in the case of heterogeneous classifiers and instability, when small changes in the initial sample lead to significant changes in the classification. To increase the speed of combining decisions of basic classifiers, a new quantum-inspired method of collective decision-making based on metaheuristic quantum algorithms is proposed. The development of ensemble methods in high-speed online learning is expected in the future.

REFERENCES

1. Friedman, J. & Greedy, H. (2001). Function approximation: a gradient boosting machine. Annals of statistics, 29(5), 1189-1232.
2. Zhou, ZH. & Wu, J. & Tang, W. (2002). Ensembling neural networks: many could be better than all. Artificial intelligence, 137(1), 239-263.
3. Sahu, A. & Runger, G. & Apley, D. (2011). Image denoising with a multi-phase kernel principal component approach and an ensemble version. IEEE applied imagery pattern recognition workshop, 1-7.
4. Shinde, A. & Sahu, A. & Apley, D. & Runger, G. (2014). Preimages for variation patterns from kernel PCA and bagging. IIE Transactions, 46(5), 429-456.
5. Buhlmann, P. & Hothorn, T. (2007). Boosting algorithms: Regularization, prediction and model fitting. Statistical Science, 477-505.
6. Arya, S. & Mount, D. & Netanyahu, N. & Silverman, R. & Wu, A. (1998). An optimal algorithm for approximate nearest neighbor searching fixed dimensions. Journal of the ACM, 45(6), 891-923.
7. Aizerman, M. & Bravermann, E. and Rosonoer, L. (1970). The Potential Function Method in Machine Learning Theory. Moscow: Science.
8. Beletskaya, S. & Asanov, Yu. & Povalyaev, A. & Gaganov, A.V. (2015). Research of the effectiveness of genetic algorithms for multicriteria optimization. Voronezh State Technical University Bulletin, 11(1), 1-4.
9. Epanechnikov, V. (1969). Nonparametric estimation of multidimensional probability density. Probability theory and its applications, 14(1), 156-161.
10. Dempster, A. & Laird, N. & Rubin, D. (1977). Maximum likelihood from incomplete data via the EM algorithm. Journal of the Royal Statistical Society. Series B (Methodological), 39(1), 1-38.
11. Scholkopft, B. & Mullert, K. & Fisher, R. (1999). Discriminant analysis with kernels. Neural Networks for Signal Processing, 1(1), 41-48.
12. Hosmer Jr, DW. & Lemeshow, S. & Sturdivant, R. (2013). Applied logistic regression. New- York: John Wiley & Sons.
13. Cortes, C. & Vapnik, VN. (1995). Support-vector networks. Machine Learning, 20(3), 273297.
14. Kamiski, B. & Jakubczyk, M. & Szufel, P. (2017). A framework for sensitivity analysis of decision trees. Central European Journal of Operations Research, 26(1), 135159.
15. Quinlan, JR. (2014). C4.5: programs for machine learning. Amsterdam: Elsevier.
16. Breiman, L. & Friedman, J. & Stone, CJ. & Olshen, RA. (1984). Classification and regression trees. Monterey: Wadsworth & Brooks.
17. Rutkovskaya, D. & Rutkovsky, L. & Pilinsky, M. (2013). Neural networks, genetic algorithms and fuzzy systems. Moscow: Hotline-Telecom.
18. Glover, F. & Kochenberger. G.A. (2003). Handbook on Metaheuristics. Kluwer Academic Publishers.

19. Veldhuizen, DAV. & Lamont, GB. (2000). Multiobjective evolutionary algorithms: Analyzing the state-of-the-art. Journal of Evolutionary Computation, The MIT Press, 8(2), 125147.

20. Fabris, F. & Krohling, RA. (2012). A co-evolutionary differential evolution algorithm for solving min-max optimization problems implemented on gpu using c-cuda. Expert Syst. Appl., 39(6), 1032410333.

21. Google llc. (2019). quantum, google ai. https://ai.google/research/teams/applied-science/quantum-ai/, 2019. Accessed: May. 18, 2021, [Online].

22. IBM. (2019). ibm q, quantum computing. https://www.research.ibm.com/ibm-q/, 2019. Accessed: May. 4, 2021. [Online].

23. Intel corporation. (2019). 2018 ces: Intel advances quantum and neuromorphic computing research. https://newsroom.intel.com/news/intel-advances-quantumneuromorphic-computing-research/,2019. Accessed: May. 2, 2021, [Online].

24. D-wave systems. (2019). dwave systems. https://www.dwavesys.com/home/, 2019. Accessed: May. 15, 2021, [Online].

25. Kirkpatrik, S. & Gelatt, CD. & Vecchi. MP. (1995). Optimization by simulated annealing. Science, 220, 671680.

26. Kennedy, K. & Eberhart, R. (1995). Particle swarm optimization. in: Proceedings of the IEEE International Conference on Neural Networks (ICNN95), Perth, Australia, 4, 19421948.

27. Storn, R. & Price, K. (1995). Differential evolution a simple and efficient heuristic for global optimization over continuous spaces. Technical Report TR-95-012, ICSI.

28. Dorigo, M. & Maniezzo, V. & Colorni, A. (1996). The ant system: optimization by a colony of cooperating agents. IEEE Trans. Syst. Man & Cybernet. Part B, 26(1), 2941.

29. Narayanan, A. & Moore, M. (1996). Quantum-inspired genetic algorithms. in: Proceedings of IEEE Int. Conf. Evol. Comput., 6166.

30. Han, KH. & Kim, JH. (2000). Genetic quantum algorithm and its application to combinatorial optimization problem. in: Proceedings of Congr. Evol. Comput.(CEC), 2, 13541360.

31. Montiel, O. & Rubio, Y. & Olvera, C. & Rivera, A. (2019). Quantum inspired acromyrmex evolutionary algorithm. Nat. Sci. Rep., 9(12181), 169176.

32. 32. Han, KH. & Park, KH. & Lee, CH. & Kim, JH. (2001). Parallel quantum inspired genetic algorithm for combinatorial optimization problem. in: Proceedings of Congr. Evol. Comput., 2, 14221429.

33. Li, Y. & Shi, H. & Jiao, L. & Liu. R. (2012). Quantum evolutionary clustering algorithm based on watershed applied to sar image segmentation. Neurocomputing, 87(10), 9098.

34. Wang, Y. & Feng, XY. & Huang, YX. & Pu, DB. & Zhou, WG. & Liang, YC. & Zhou, CG. (2007). A novel quantum swarm evolutionary algorithm and its applications. Neurocomputing, 70(4), 633640.

35. Zouache, D. & Nouioua, F. & Moussaoui, A. (2016). Quantum inspired firefly algorithm with particle swarm optimization for discrete optimization problems. Soft Computing, 20(7), 27812799.

36. Zhang, G. & Gheorghe, M. & Wu. C. (2008). A quantum-inspired evolutionary algorithm based on p systems for knapsack problem. Fundam. Inform., 87(1), 93116.

37. Zouache, D. & Moussaoui, V. (2015). Quantum-inspired differential evolution with particle swarm optimization for knapsack problem. J. Inf. Sci. Eng., 31, 17791795.

38. Meng, K. & Wang, HG. & Dong, Z. & Wong, KP. (2010). Quantum inspired particle swarm optimization for valve-point economic load dispatch. IEEE Transactions on Power Systems, 25(1), 215222.

39. Tehzeeb ul Hassan, H. & Asghar, MU. & Zamir, MZ. & Faiz, HMA. (2017). Economic load dispatch using novel bat algorithm with quantum and mechanical behaviour. In: Proceedings of 2017 International Symposium on Wireless Systems and Networks (ISWSN), 16.

40. Ullah, S. & and Wahid, & M. (2015). Topology control of wireless sensor network using quantum inspired genetic algorithm. International Journal of Swarm Intelligence and Evolutionary Computation, 04, 08.

41. Zhang, G. & Zhang, G. & Gao, Y. & Lu, J. (2011). Competitive strategic bidding optimization in electricity markets using bilevel programming and swarm technique. IEEE Transactions on Industrial Electronics, 58(6), 21382146.

42. Dey, S. & Bhattacharyya, S. & Maulik, U. (2014). Quantum inspired genetic algorithm and particle swarm optimization using chaotic map model based interference for gray level image thresholding. Swarm and Evolutionary Computation, 15, 3857.

43. Dey, S. & Bhattacharyya, S. & Maulik, U. (2017). Quantum-inspired multi-objective simulated annealing for bilevel image thresholding. Quantum Inspired Computational Intelligence, Research and Applications.

44. Dey, S. & Bhattacharyya, S. & Maulik, U. (2013). Quantum inspired meta-heuristic algorithms for multi-level thresholding for true colour images. In: Proceeding of 2013 Annual IEEE India Conference (INDICON), Mumbai, India.

45. Tkachuk. V. (2018). Quantum genetic algorithm on multilevel quantum systems. Mathematical Problems in Engineering, 112.

46. Dey, S. & Bhattacharyya, S. & Maulik, U. (2016). New quantum inspired meta-heuristic techniques for multi-level colour image thresholding. Applied Soft Computing, 46, 677702.

47. 47. Dey, S. & Bhattacharyya, S. & Maulik, U. (2017). Efficient quantum inspired meta-heuristics for multi-level true colour image thresholding. Applied Soft Computing, 56, 472513.

48. Ctardenas-Ltopez, FA. & Lamata, L. & Retamal, JC. & Solano, E. (2018). Multiqubit and multilevel quantum reinforcement learning with quantum technologies. PLoS ONE, 13(7), 112.

49. Dey, S. & De, S. & Ghosh, D. & Konar, D. & Bhattacharyya, S. & Platos, J. (2019). A novel quantum inspired sperm whale metaheuristic for image thresholding. In: Proceedings of 2019 Second International Conference on Advanced Computational and Communication Paradigms (ICACCP), 17.

50. Dutta, T. & Bhattacharyya, S. & Dey, S. & Platos, S. (2020). Border Collie Optimization. IEEE Access, 8, 109177-109197.

51. Kim, Y. & Kim, JH. & Han, KH. (2006). Quantum-inspired multiobjective evolutionary algorithm for multiobjective 0/1 knapsack problems. In: Proceedings of IEEE Int. Conf. Evol. Comput., 26012606.

52. Moghadam, MS. & Nezamabadi-Pour, H. & Farsangi, MM. (2005). A quantum behaved gravitational search algorithm. Intell. Inf. Manage., 2012(4), 390395.

53. Chakraborti, T. & Chatterjee, A. & Halder, T. & Konar. A. (2015). Automated emotion recognition employing a novel modified binary quantum-behaved gravitational search algorithm with differential mutation. Expert Syst., 32(4), 522530.

54. Li, B. & Wang, L. (2007). A hybrid quantum-inspired genetic algorithm for multiobjective flow shop scheduling. IEEE Trans. Syst., Man, Cybern. B, Cybern., 37(3), 576591.

3 Function Optimization Using IBM Q

3.1 INTRODUCTION

An optimization problem is the searching technique to find out the best solution from all the possible solutions. When the practical optimization problem in any discipline is represented by a mathematical function, it is called objective function. Real-world optimization problems are typically having several objective functions. There are many ways available in literature for resolving the practical optimization problems. In this chapter, the extensively used optimization techniques are described briefly. The main focus has been given to solve the optimization problems using IBM Q. After introduction in the first section of this chapter, a brief overview to single-objective and multi-objective optimization with their difficulties have been discussed.

Section 3.3 has been reserved to discuss about modern techniques of resolving optimization problems. Genetic algorithms, simulated annealing, particle swarm optimization, differential evolution, ant colony optimization, bee-colony optimization, harmony search algorithm, bat-algorithm, Cuckoo search, neural network-based optimization, fuzzy optimization, etc. are available in literature, which are considered as modern methods of optimization problem solving. All the afore-mentioned methods are described in short with their consequences.

When the complexity of optimization problems and amount of data involved rise, more efficient ways of solving optimization problems are needed. The power of quantum computing can be used for solving problems which are not practically feasible on classical computers, or suggest a considerable speed up with respect to the best known classical algorithm. In Section 3.4, the author has enlightened on the basics of quantum computing and the quantum algorithms used for solving optimization problems.

IBM provides an experimental cloud-enabled quantum computing user-interface platform, known as IBM Q, for the students, researchers, and general science enthusiasts. It allows users to run established algorithms and experiments, work with quantum bits (qubits), etc. The different features of IBM Q have been reflected in Section 3.5.

The circuit composer in IBM Q is a tool that allows users to visually learn how to create quantum circuits. In Section 3.6, the step-by-step approach has been presented to build a sample quantum application using circuit composer.

Qiskit (Quantum Information Software Kit) in IBM Q is an open-source quantum computing framework, which enables developers and researchers to conduct quantum explorations using Python scripts. Section 3.7 has been utilized to showcase a

DOI: 10.1201/9781003283294-3

sample Qiskit project with its different components like Qiskit Terra, Qiskit Aqua, Qiskit Ignis, and Qiskit Aer.

In Section 3.8, the author has enlightened on the application of quantum computing, where IBM Q can be utilized seamlessly. Out of those applications, objective function optimization using IBM Q has been broadly illustrated in this section. Portfolio optimization, risk analysis, and Monte-Carlo-like applications are considered as few examples of optimization.

The last section of this book chapter is dedicated for drawing a conclusionary communication in terms of the application of IBM Q in objective function optimization.

3.2 FUNCTION OPTIMIZATION

The main reason to rely on any model in a decision-making process is to provide a quantitative assessment of the effects of management decisions on the system being considered. A model also provides a fairly objective assessment as opposed to subjective opinions of system behavior. Thus, models should be used in support of decision-making. Optimization is just another type of modeling and the same applies to optimization models. Optimization tools should be used for supporting decisions rather than for making decisions, that is, should not substitute decision-making process.

3.2.1 DIFFICULTIES IN OPTIMIZATION METHODS

Real-world optimization problems often involve a number of characteristics, which make them difficult to solve up to a required level of satisfaction. The characteristics are mentioned below [1]

1. Existence of mixed type of variables (such as Boolean, discrete, integer, and real)
2. Existence of non-linear constraints
3. Existence of multiple conflicting objectives
4. Existence of multiple optimum (local and global) solutions
5. Existence of stochasticity and uncertainty in describing the optimization problem

Most real-world search and optimization problems involve multiple objectives. A solution that is optimum (maximum or minimum) with respect to one objective may not be optimum with respect to another objective. So some compromise must be required. This is more difficult to implement multi-objective optimization problems (MOOPs) than single objective optimization problems (SOOPs).

3.2.2 DEFINITION OF MULTI-OBJECTIVE OPTIMIZATION PROBLEM (MOOP)

A multi-objective optimization problem (MOOP) has a number of objective functions, which are to be minimized or maximized. Multi-objective optimization

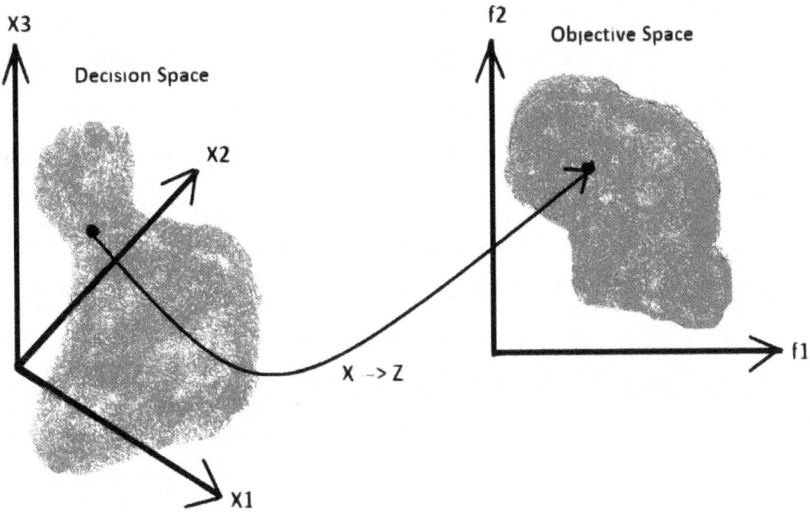

Figure 3.1: Decision variable space and the corresponding objective space.

problem can be described in its general form:

$$
\begin{aligned}
Minimize/Maximize \quad & f_m(x), & m &= 1,2,\ldots,M; \\
Subject\ to \quad & g_j(x) >= 0, & j &= 1,2,\ldots,J; \\
& h_k(x) = 0, & k &= 1,2,\ldots,K; \\
& x_i^{(L)} <= x_i <= x_i^{(U)}, & i &= 1,2,\ldots,n;
\end{aligned}
\tag{3.1}
$$

As indicated in Figure 3.1, a solution "x" is a vector of "n" decision variables: $x = (x_1, x_2, \ldots, x_n)^T$. The last set of constraints is called variable bounds. These bounds constitute a decision variable space, D. The terms "$g_j(x)$" and "$h_k(x)$" are called constraint functions. In this problem, there are 'J' inequality (greater-than-equal-to type) and "K" equality constraints. A solution "x" that does not satisfy all of the $(J + K)$ constraints and all of the variable bounds is called an infeasible solution. On the other hand, if any solution "x" satisfies all constraints and variable bounds is known as feasible solution. So, in the presence of constraints, the entire decision variable space D may not be feasible. The set of all feasible solutions is called the feasible region, S. For each solution "x" in the decision variable space, there exists a corresponding point in another space, which is called objective space, Z. The mapping takes place between an n-dimensional solution vector and an M-dimensional objective vector[1].

3.2.3 DIFFERENCES BETWEEN SOOPS AND MOOPS

The main goal of single-objective optimization is to find the best solution, which corresponds to the minimum or maximum value of a single objective function that lumps all different objectives into one. This type of optimization is a useful tool, which should provide decision makers with insights into the nature of the problem, but usually cannot provide a set of alternative solutions that trade different objectives against each other. On the contrary, in a multi-objective optimization with conflicting objectives, there is no single optimal solution. The interaction among different objectives gives rise to a set of compromised solutions, largely known as the trade-off, non-dominated, noninferior, or Pareto-optimal solutions.

1. In a single-objective optimization, there is only one goal, the search for an optimum solution. In multi-objective optimization, there are two goals, progressing toward the Pareto-optimal front and maintaining diversity among the solutions in the Pareto-optimal front.
2. In a single-objective optimization, there is only one search space, the decision space. But in multi-objective optimization, there are two search spaces, decision space and objective space.
3. Single-objective optimization is the degenerate case of multi-objective optimization. In many cases, multi-objective optimization can be converted into single-objective optimization.

3.3 MODERN OPTIMIZATION PROBLEM-SOLVING TECHNIQUES

Heuristic and metaheuristic are the two common terms while considering modern optimization techniques to solve the optimization problems. Heuristic means to find or discover certain value or solution by trial and error. Heuristic algorithms can attain to the optimum solutions within a given finite time, but it never assures its performance in achieving the best solution. To achieve more better and accurate solutions, meta-heuristic algorithms have been developed by making some improvements over heuristic algorithms. The modern metaheuristic algorithms work on the basis of certain randomization of the search process. In the subsequent subsections, a number of modern optimization problem-solving techniques have been discussed briefly with their limitations. These methods operate by considering the issues such as imprecision, partial truth, uncertainty, and approximation, which are often present in few real-world problems.

3.3.1 GENETIC ALGORITHM

Genetic Algorithm (GA), which mimics the natural process of Evolution and Darwin's principle of "Survival of the Fittest", was invented by John Holland in the 1960s at the University of Michigan. According to [7], GAs are particularly suitable for solving complex optimization problems like wire routing, scheduling, adaptive control, game playing, cognitive modeling, transportation problems, traveling salesman problems, optimal control problems, database query optimization, etc.

A GA comprises a set of individual elements (the population) and a set of bio-logically inspired operators defined over the population itself. According to evolu-tionary theories, only the most suited elements in a population are likely to survive and generate offspring, thus transmitting their biological heredity to new genera-tions. In computing terms, a genetic algorithm maps a problem onto a set of strings, each string representing a potential solution. A simple GA cycle has four stages: population of strings, evaluation of each string, selection of best string, and genetic operators to create the new population of strings.

The biggest limitation of GA is that it cannot guarantee optimality. The solution quality also deteriorates with the increase of problem size.

3.3.2 SIMULATED ANNEALING

Simulated Annealing (SA) uses a thermodynamic evolution process to search min-imum energy states [3]. The primary objective of simulated annealing is to find the global minimum of a function that characterizes large and complex systems. It pro-vides a powerful tool for solving non-convex optimization problems. Simulated an-nealing starts from a random initial solution. It then proceeds by generating new solution and accepting/rejecting them probabilistically. As the search proceeds, the temperature cools down and the process converges to a global minima. The tempera-ture needs to be reduced at a slow and controlled rate to ensure proper solidification with a low energy crystalline state that corresponds to the best required result.

If the cost function is expensive to compute, the algorithm becomes very slow. For problems where the energy landscape is smooth, or there are few local minima, SA is overkill. The method cannot tell whether it has found an optimal solution. Some other method is required to do this.

3.3.3 PARTICLE SWARM OPTIMIZATION

As per [8], Particle Swarm Optimization (PSO) algorithm was inspired from the behavior of swarm of insects, such as ants, bees, wasps, or a school of fish or a flock of birds. The algorithm operates through the behavioral intelligence of each solution/particle or insect in a swarm/ function.

Ant Colony Optimization (ACO) is based on the cooperative behavior of real ant colonies, that is, their characteristics of finding the shortest path from their nest to a food source.

Bee Colony Optimization (BCO) works mainly on the basis of foraging behavior of the bees. The natures of bee, that is, forage which means hunt or explore, are taken consideration for this optimization algorithm.

Like GA, PSO algorithms do also not guarantee about the optimum solution. Al-though ACO and BCO have higher probability and efficiency in finding the global optima, it is difficult to define their initial design parameters and not suitable for large-scale problems.

3.3.4   BAT ALGORITHM

Bats emit a loud sound pulse and listen for the echo that reflects from adjacent objects. Depending on this echo voice, they avoid the obstructions and treasure their root while searching their target. Based on certain properties of their pulses emitted and the time for detecting the echo by their ears, the bats are able to detect the distance, orientation, type of victims, etc. This optimization algorithm [5], utilizes the echolocation nature of bats.

Bat algorithm converges very quickly at the early stage and there after convergence rate slows down.

3.3.5   CUCKOO SEARCH ALGORITHM

Cuckoo Search (CS) algorithm [6] is inspired by the obligate brood parasitism of some cuckoo species by laying their eggs in the nests of other host birds of different species. Female cuckoos can lay their eggs very similar (in color and pattern of eggs) to the host so that the host can take responsibility of those eggs.

CS has its own limitation while performing for multi-objective problems and discrete problems.

3.3.6   FUZZY SYSTEM

Most natural language is fuzzy, which involves vagueness and imprecision. A fuzzy logic proposition is a linguistic statement relating some perception without clearly defined boundaries. In other words, all truths in fuzzy logic are partial or approximate. Linguistic statements that tend to express subjective ideas and that can be interpreted slightly differently by various individuals typically involve fuzzy propositions. Fuzzy propositions are assigned to fuzzy sets. The objective and constraint functions are characterized by the membership functions in a fuzzy system, where membership in a classical subset, of a classical crisp sets of objects.

Fuzzy logic-based control systems, i.e. Fuzzy Logic Controller (FLC), can handle many real-world problems that cannot be efficiently managed by conventional (crisp) control systems. Fuzzy control systems theory uses the routine fuzzification fuzzy operation defuzzification [7]. The purpose of designing and applying FLCs is to tackle vague, ill-described, and complex processes that can barely be handled by classical systems theory, classical control techniques, and classical two-valued logic. The majority of FLCs are knowledge-based systems in that either their fuzzy models or their fuzzy logic controllers are described by fuzzy IF-THEN rules. Rules are established based on experts' knowledge about the systems, controllers, performance, etc.

Fuzzy system is also having the limitations. The main concern is stability of the system. Determining the fuzzy rules and membership functions are not very easy task to achieve.

3.3.7 NEURAL NETWORK-BASED OPTIMIZATION

An Artificial Neural Network (ANN) is an information processing system that has several performance features in common with biological neurons. An ANN is characterized by (a) its design pattern between the neurons, called its architecture, (b) its technique of determining the weights on the connections, called its learning algorithm, and (c) its activation functions. ANN with its different architectures and learning mechanism is suitable for solving optimization problems.

However, a few limitations of this technique are unexpected behavior of the network, determination of the proper network structure, hardware dependency, etc.

In many cases, good results have been achieved by combining the above-mentioned modern optimization problem-solving methods. Neuro-fuzzy architecture is a very good example in which the good properties of both neural network and fuzzy logic are attempted to bring together. Neuro-fuzzy systems are mostly fuzzy rule-based systems, in which different techniques of neural networks are used for rule introduction and calibration. Fuzzy logic may also be employed to improve the performance of optimization methods used with neural networks [8].

3.4 QUANTUM COMPUTING AND OPTIMIZATION ALGORITHMS

The common limitation of the above-mentioned modern optimization problem-solving techniques is convergence speed. This problem has been resolved by using quantum computing, which is much faster than the traditional computing. This section is dedicated for a brief overview of quantum computing and optimization algorithms.

3.4.1 QUANTUM COMPUTING

All computing systems rely on a fundamental ability to store and manipulate information. Current computers manipulate individual bits, which store information as binary 0 and 1 states. Quantum computers leverage quantum mechanical phenomena to manipulate information. To do this, they rely on quantum bits, or qubits, depicted in IBMs quantum computing reference website. Three quantum mechanical properties (superposition, entanglement, and interference) are used in quantum computing to manipulate the state of a qubit.

The fundamental building block, i.e. Quantum bit or Qubit, of quantum computer has been represented by Bloch sphere in Figure 3.2. A qubit can be at 1 state or 0 state or in any superposition of the two. The state of a qubit can be represented as

$$\Psi\rangle = \alpha|0\rangle + \beta|1\rangle \tag{3.2}$$

where α and β are complex numbers and $|\alpha|^2 + |\beta|^2 = 1$

Quantum logic gates, operating on a number of qubits, are the building blocks of quantum circuits. There are many types of quantum gates, like H (Hadamard) gate, CX (Controlled-X) gate, ID (Identity) gate, U3 gate, U2 gate, U1 gate, Rx gate, Ry gate, Rz gate, X gate, Y gate, Z gate, S gate, Sdg gate T gate, Tdg gate, cH gate, cY

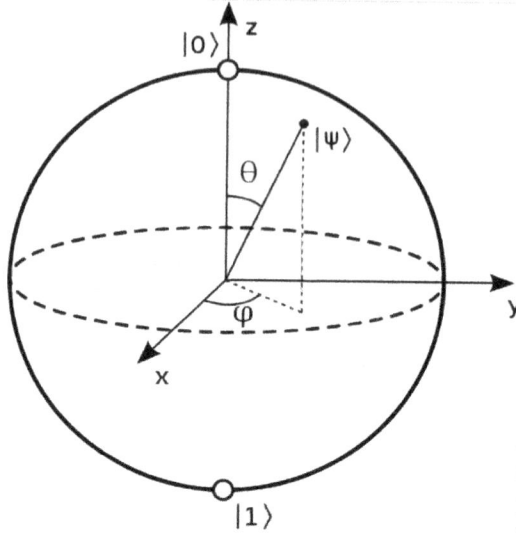

Figure 3.2: Bloch sphere.

gate, cZ gate, cRz gate, cU1 gate, cU3 gate, ccX gate, SWAP gate, etc. Even custom gates can also be created for using in quantum circuits.

Digital quantum computers use quantum logic gates to do computation. A quantum computer consists of the below-mentioned blocks or chambers, IBMs quantum computing reference website [9]

(a) Qubit Signal Amplifier
(b) Input Microwave Lines
(c) Superconducting Coaxial Lines
(d) Cryogenic Isolators
(e) Quantum Amplifiers
(f) Cryoperm Shield
(g) Mixing Chamber

3.4.2 OPTIMIZATION USING QUANTUM COMPUTING

The limitations of the individual modern optimization problem-solving techniques highlighted in Section 3.3, have been tried to minimize by using quantum computing. In general, logistics or supply chain network infrastructure, air traffic control work scheduling, and financial services are few optimization examples, which can be resolved by quantum computing.

In [10], a novel Genetic Quantum Algorithm (GQA) is introduced to resolve the optimization problem. GQA has an outstanding capability of global search due to its diversity produced by the probabilistic representation. Well-known Knapsack problem has been used to discuss the performance of GQA. Further improvement to this algorithm had been made using parallelism feature.

Quantum annealing algorithm such as Quantum Processing Unit (QPU) is applicable for solving binary optimization problems [11]. In [12], quantum annealer has been utilized to optimize the traffic flow, as mentioned by [13]. The QPU is designed to solve Quadratic Unconstrained Binary Optimization (QUBO) problems, where each qubit represents a variable and couplers between qubits represent the costs associated with qubit pairs. Quantum annealing algorithm has been used in [14] to resolve the Nurse Scheduling Problem (NSP), which arises when searching the optimal schedule for a set of available nurses to create a rotating roaster.

Quantum Adiabatic Algorithm (QAA) [15] has been used on a quantum computer for finding the global minima of a classical cost function. Performance have been measured in [16] by generating over 200,000 instances of MAX 2-SAT on 20 qubits.

Few real-life optimization examples, which can be resolved by quantum computing are

a. Telecommunications companies upgrading their network infrastructure
b. Healthcare firms optimizing patient treatments
c. Governments improving air traffic control
d. Consumer products and retail companies tailoring marketing offers
e. Financial services firms enhancing their risk optimization
f. Organizations developing employee work schedules
g. Universities scheduling classes

3.5   FEATURES OF IBM Q EXPERIENCE

The IBM Q Experience is an online platform available in Cloud, which gives users access to a set of IBM's prototype quantum processors. It was launched by IBM in May 2016. IBM Q is an industry first initiative to build universal quantum computers for business, engineering, and science. This effort includes advancing the entire quantum computing technology stack and exploring applications to make quantum broadly usable and accessible. The snapshots of IBM Q have been demonstrated in **Figure 3.3**. IBM Q is applicable to solve the most challenging problems in chemistry, optimization, machine learning, finance, etc. Many self-explanatory documentations are provided in IBMs website on quantum computing.

Official website - https://quantum-computing.ibm.com/

3.6   CIRCUIT COMPOSER IBM Q

The circuit composer is a tool in IBM Q that allows users to visually create quantum circuits. **Figure 3.4** is the visual representation of circuit composer in IBM Q.

Figure 3.5 shows a sample circuit diagram in visual form and in internal coding form for 5 quantum registers. The circuit can be exported as .qasm file, which can be again modified and imported. **Figure 3.6** shows the result dashboard after running the circuit.

Figure 3.3: Snapshot of IBM Q.

Circuit composer of IBM Q is being widely used to simulate the quantum circuits visually and test those very easily. A quantum calculator (addition, subtraction, multiplication, and division) is demonstrated and simulated using circuit composer [17]. The algorithm for obtaining maximum and minimum of any mathematical model has been presented in [18]. Circuit composer has been used as a simulator to find out the minimum from Titanic passengers age. In [19], controlled square root of Z gate has been constructed and tested using circuit composer. An exhaustive survey

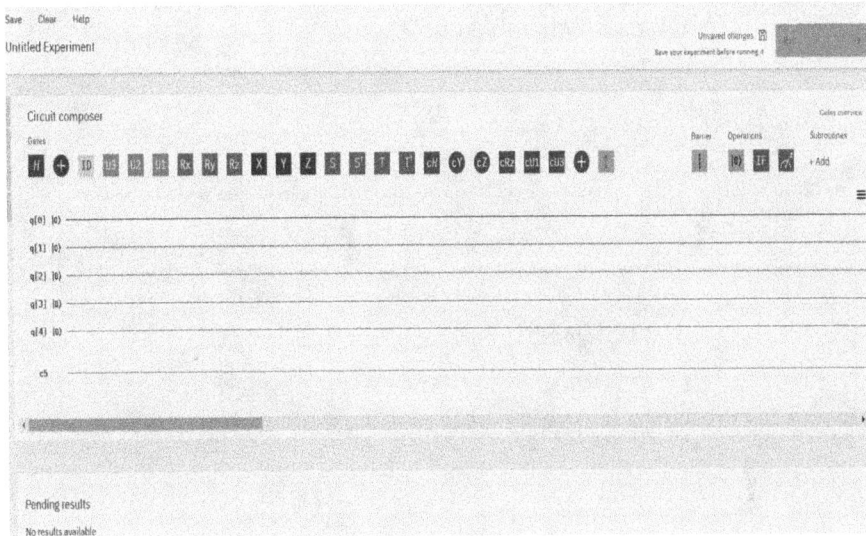

Figure 3.4: Circuit composer.

on 20 different quantum algorithms is available in literature [20]. All the algorithms are simulated and tested using circuit composer.

3.7 QISKIT IN IBM Q

QISKit [21] is an open-source quantum computing software development kit for leveraging today's quantum processors in research, education, and business. It is a Python-based software library that can be used to create quantum computing programs, compile, and execute them on one of several backends [22]. So, QISKit can be installed on top of Python using the below command *pip install qiskit* This is also available in IBM Q Experience with the Qiskit Notebooks as shown in **Figure 3.7**.

3.7.1 CREATING 5-QUBIT CIRCUIT WITH THE HELP OF QISKIT IN IBM Q

Once the New Notebook button is clicked, a Jupyter notebook gets opened with QISKit installed into it. After writing the notebook completely, it can be saved as .ipynb. This notebook can be downloaded as different formats as shown in **Figure 3.8**.

A sample QISKit code has been written to create a quantum circuit using a Hadamard gate H on qubit 0, a Controlled-X gate on control qubit 0 and target qubit 1 and measure on each qubit to visualize the result. In **Figure 3.9**, a sample circuit has been drawn using QISKit programming and **Figure 3.10** shows simulation and plotting the result.

Circuit diagram

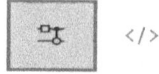

Original circuit diagram

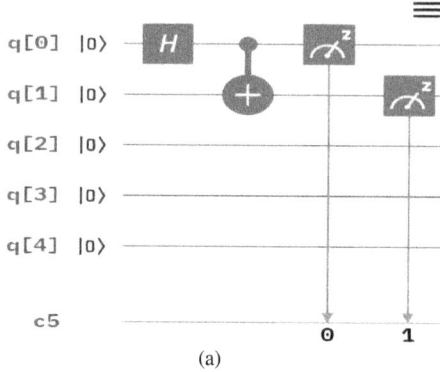

q[0] |0⟩ ──[H]──●──[⤢ᶻ]────────

q[1] |0⟩ ─────────⊕──────[⤢ᶻ]──

q[2] |0⟩ ─────────────────────

q[3] |0⟩ ─────────────────────

q[4] |0⟩ ─────────────────────

c5   ─────────────────0───1──

(a)

Circuit diagram

Original circuit diagram

```
1     OPENQASM 2.0;
2     include "qelib1.inc";
3
4     qreg q[5];
5     creg c[5];
6
7     h q[0];
8     cx q[0],q[1];
9     measure q[0] -> c[0];
10    measure q[1] -> c[1];
```

(b)

Figure 3.5: (a) Diagrammatic representation, (b) Pseudo-code representation.

3.7.2   TESTING THE CIRCUIT USING IBM QUANTUM COMPUTER

Up to this point, QISKit codes were running on Jupyter notebook locally. Now, this data has been sent to IBM Q machine ibm‵16‵melbourne, which has been shown in **Figure 3.11**. It is treated as a job, which has different statuses like

a. Job is validating
b. Job is successfully queued
c. Job has successfully run

Once job has successfully run, the results can be plotted using plot‵histogram function.

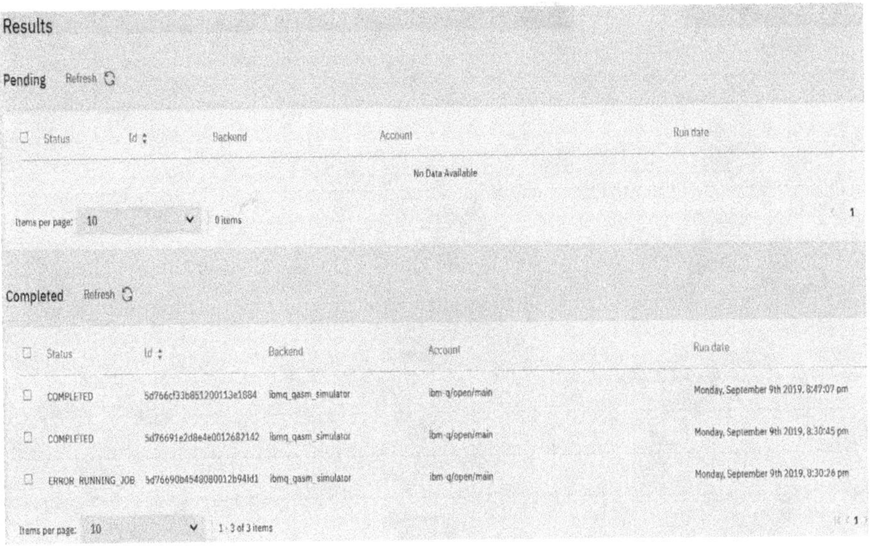

Figure 3.6: Result dashboard.

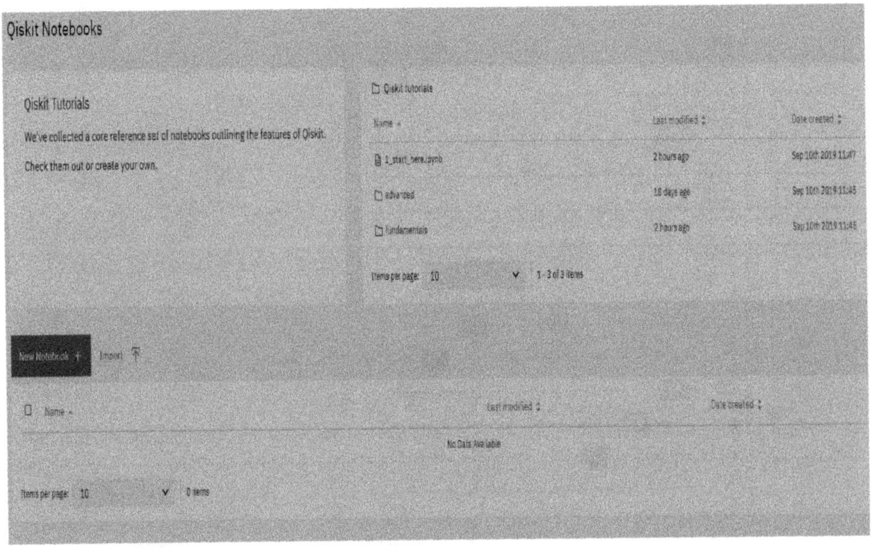

Figure 3.7: QISKit notebooks.

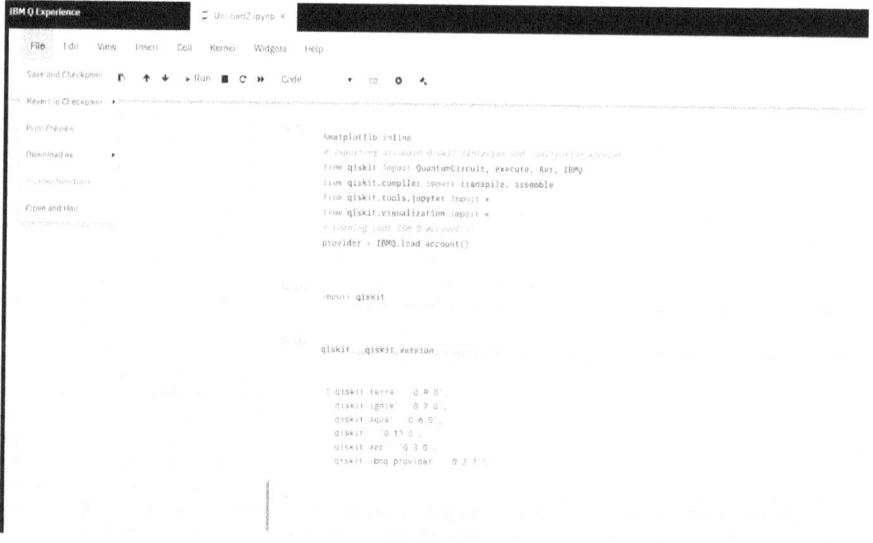

Figure 3.8: Working with QISKit notebook.

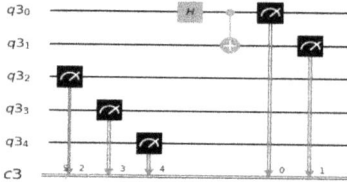

Figure 3.9: Sample 5 qubit circuit using QISKit.

```
In [11]:   # Simulate the experiment
           simulator = Aer.get_backend('qasm_simulator')
           result = execute(circuit,backend=simulator).result()
           hist = result.get_counts(circuit)
           print("\nTotal count for 00 and 11 are:",hist)
           # Visualize the results
           plot_histogram(hist)
```

Total count for 00 and 11 are: {'00000': 505, '00011': 519}

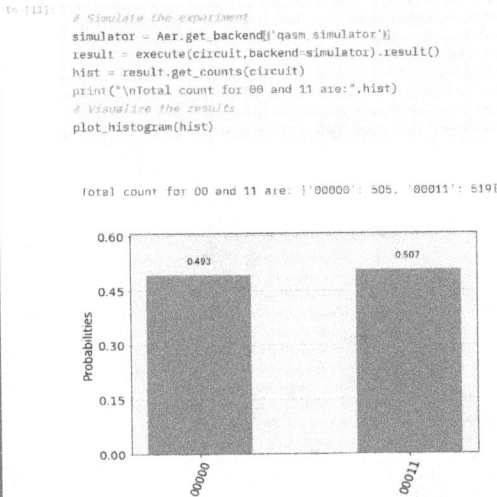

Figure 3.10: Simulating and plotting the result.

From the two plots provided in **Figure 3.10** and **Figure 3.11**, the differences are very clear. When the result on Jupyter notebook is run, probability of occurrences is ideal and caught at 2 significant qubits 00000 and 00011. The probability of occurrences in other qubits is ignored. When the result is sent to the real quantum computer, that is, IBM Q for analysis, it provides probability of occurrences in many qubits by considering the minimal errors too. By improving the technology of real quantum computer, the scientists are working out to diminish these insignificant errors.

print(result) = '00011': 513, '00000': 511

print(result'ibmq) = '00101': 1, '01011': 5, '00010': 36, '10010': 2, '01010': 1, '10000': 20, '00011': 311, '00111': 2, '00001': 54, '00000': 571, '01000': 4, '10001': 1, '11111': 1, '11011': 1, '00100': 2, '10011': 11, '01001': 1

3.8 OPTIMIZATION USING IBM Q

One of the application areas of IBM Q is to resolve the most challenging optimization problems. Quantum computers offer the best or optimal solution among varying weighted selections more competently than classical computers. Qiskit Aqua [23] is a tool, which allows researchers to experiment with optimization applications. It contains a library of cross-domain quantum algorithms upon which applications for near-term quantum computing can be built. The Aqua algorithms run on top of Qiskit Terra, which builds, compiles, and executes quantum circuits on simulators or real quantum devices.

```
In [14]: # Sending data to live IBM Q machine and monitoring job
         provider = IBMQ.get_provider('ibm-q')
         qcomp = provider.get_backend('ibmq_16_melbourne')
         job = execute(circuit, backend=qcomp)
         from qiskit.tools.monitor import job_monitor
         job_monitor(job)

         Job Status: job has successfully run

In [15]: # Visualize the results
         result2 = job.result()
         hist2 = result2.get_counts(circuit)
         plot_histogram(hist2)
```

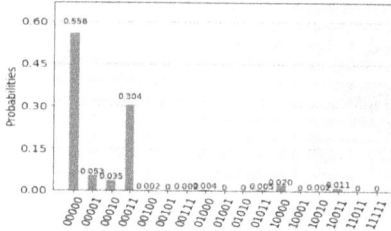

Figure 3.11: Creating a job for IBM Q and visualizing the result.

Travelling Salesman Problem (TSP), considered as NP-hard problem, is an important category of optimization problems that is mostly encountered in various areas of science and engineering. In [24], Circuit Composer of IBM Q has been used to simulate the TSP by considering four cities.

Vehicle Routing Problem (VRP), another NP-hard problem, comprises of searching the smallest distance for delivering goods from a depot to many geographically distributed locations with numerous journeys. VRP has been considered to solve using IBM Q Experience and presented in [25].

The classical Tabu Search is a metaheuristic that explores search spaces and conducts a local heuristic search procedure to explore the solution space beyond local optimum using a Tabu list with forbidden moves. In [26], the Quantum Tabu Search (QTS) algorithm has been proposed based on Knapsack problem approach. The proposed QTS has been simulated and tested by using backend IBM Q 16 Rueschlikon (16-qubits) and IBM Q 5 Yorktown (5-qubits) simulator.

The number of gates and the number of levels used in IBM Q have to be optimized to obtain the better result. This improves the reliability of the output state, reduces the effect of noise, and increases the accuracy of the quantum computation. In [27], circuits implemented in IBMs QX2 and QX4 have been utilized with maximum 16 qubits to establish experimental violation of Mermin inequality.

Variational Quantum Eigensolver (VQE) [28] is a hybrid between classical and quantum computing. A classical computer controls the preparation of a quantum state using few experimental parameters; then a quantum computer prepares that state and calculates its properties. Optimization problems can be resolved by quantum computer using VQE. In [29], VQE in IBM Q has been applied to solve the MaxCut NP-complete binary optimization problem with 5 qubits.

Grover's Adaptive Search (GAS) [30] is a fast quantum mechanical algorithm for combinatorial optimization problems, which can resolve an O(N/2) optimization problem into O(N) steps. In [31], a modified version of Grover's search algorithm with fewer gates, optimized number of iterations and improved performance has been presented. To establish the upgraded and optimized quantum search, set search and array search algorithms have been implemented using IBM Q. Grover optimizer can be easily applied to solve the QUBO problem [32].

In July 2020, IBM came up with Qiskit Optimization Module [33]. This is now in an initial stage by keeping in mind the goal of providing a super-optimized solution to the users within few milliseconds for any input problem. This module will act as a black box with the combination of quantum and classical resources. Users do not need to have the knowledge of quantum theory and mechanics. According to the documentation, it empowers easy and efficient modeling of optimization problems for developers and optimization experts without quantum expertise. IBMs DOcplex (Decision Optimization CPLEX) [34] modeling for python is used to develop the Qiskit Optimization Module.

3.9   CONCLUSION

As per definition, optimization is the technique of finding an alternative with the most cost-effective or highest achievable performance under the given constraints, by maximizing desired factors and minimizing undesired ones. Since many years, lot of efforts have been incorporated to resolve the optimization problems, specifically the NP-hard and multi-objective problems. It becomes slight easy after the invention of quantum computer, which is very costly and sensitive. IBM makes it available freely for the common people by introducing IBM Q Experience online platform.

In this chapter, efforts have been made to discuss the two features, circuit composer and QISkit, available in IBM Q. A 5-qubit circuit has been created on Jupyter notebook using QISKit software development kit and the same has been tested using real IBM quantum computer. The chapter has been concluded by discussing on the techniques to resolve optimization problems using IBM Q.

ACKNOWLEDGMENTS

This work had been accomplished on IBM Cloud to leverage its IBM Quantum experience platform. The author is grateful to IBM for providing access to its IBM Q environment.

REFERENCES

1. Kalyanmoy Deb. Optimization for engineering design: Algorithms and examples. PHI Learning Pvt. Ltd., 2012.

2. David E. Goldberg and John Henry Holland. Genetic algorithms and machine learning. Machine Learning, 3: 95–99, 1988.

3. Scott Kirkpatrick, C Daniel Gelatt, and Mario P Vecchi. Optimization by simulated annealing. Science, 220(4598):671–680, 1983.

4. James Kennedy and Russell Eberhart. Particle swarm optimization. In Proceedings of ICNN'95-International Conference on Neural Networks, volume 4, pages 1942–1948. IEEE, 1995.

5. Xin-She Yang. Bat algorithm for multi-objective optimisation. International Journal of Bio-Inspired Computation, 3(5):267–274, 2011.

6. Xin-She Yang and Suash Deb. Cuckoo search via Levy flights. In 2009 World Congress on Nature & Biologically Inspired Computing (NaBIC), pages 210–214. IEEE, 2009.

7. Guanrong Chen, Trung Tat Pham, and N.M. Boustany. Introduction to fuzzy sets, fuzzy logic, and fuzzy control systems. Applied Mechanics Reviews, 54(6):B102–B103, 2001.

8. Detlef Nauck, Frank Klawonn, and Rudolf Kruse. Foundations of neuro-fuzzy systems. John Wiley & Sons, Inc., 1997.

9. https://www.ibm.com/quantum-computing/.

10. Kuk-Hyun Han and Jong-Hwan Kim. Genetic quantum algorithm and its application to combinatorial optimization problem. In Proceedings of the 2000 Congress on Evolutionary Computation. CEC00 (Cat. No. 00TH8512), volume 2, pages 1354–1360. IEEE, 2000.

11. Kuk-Hyun Han, Kui-Hong Park, Ci-Ho Lee, and Jong-Hwan Kim. Parallel quantum-inspired genetic algorithm for combinatorial optimization problem. In Proceedings of the 2001 Congress on Evolutionary Computation (IEEE Cat. No. 01TH8546), volume 2, pages 1422–1429. IEEE, 2001.

12. Florian Neukart, Gabriele Compostella, Christian Seidel, David Von Dollen, Sheir Yarkoni, and B. Parney. Traffic flow optimization using a quantum annealer. Frontiers in ICT, 4:29, 2017.

13. Mark W Johnson, Mohammad HS Amin, Suzanne Gildert, Trevor Lanting, Firas Hamze, Neil Dickson, Richard Harris, Andrew J Berkley, Jan Johansson, Paul Bunyk, et al. Quantum annealing with manufactured spins. Nature, 473(7346):194–198, 2011.

14. Kazuki Ikeda, Yuma Nakamura, and Travis S Humble. Application of quantum annealing to nurse scheduling problem. Scientific Reports, 9(1):1–10, 2019.

15. Edward Farhi, Je rey Goldstone, Sam Gutmann, and Michael Sipser. Quantum computation by adiabatic evolution. arXiv preprint quant-ph/0001106, 2000.

16. Elizabeth Crosson, Edward Farhi, Cedric Yen-Yu Lin, Han-Hsuan Lin, and Peter Shor. Different strategies for optimization using the quantum adiabatic algorithm. arXiv preprint arXiv:1401.7320, 2014.

17. Prathamesh P Ratnaparkhi and K. Bikash. Demonstration of a quantum calculator on IBM quantum experience platform, DOI: 10.13140/RG.2.2.12661.63209 (2018)

18. Yanhu Chen, Shijie Wei, Xiong Gao, Cen Wang, Jian Wu, and Hongxiang Guo. An optimized quantum maximum or minimum searching algorithm and its circuits. arXiv preprint arXiv:1908.07943, 2019.

19. Petar Nikolov and Vassil Galabov. Experimental realization of controlled square root of z gate using ibm's cloud quantum experience platform. arXiv preprint arXiv:1806.02575, 2018.

20. J. Abhijith, Adetokunbo Adedoyin, John Ambrosiano, Petr Anisimov, Andreas artschi, William Casper, Gopinath Chennupati, Carleton Corin, Hristo Djidjev, David Gunter, et al. Quantum algorithm implementations for beginners. arXiv e-prints, pages arXiv:1804, 2018.

21. https://qiskit.org/.
22. Robert Wille, Rod Van Meter, and Yehuda Naveh. Ibm's qiskit tool chain: Working with and developing for real quantum computers. In 2019 Design, Automation & Test in Europe Conference & Exhibition (DATE), pages 1234–1240. IEEE, 2019.
23. https://github.com/Qiskit/qiskit-aqua.
24. Karthik Srinivasan, Saipriya Satyajit, Bikash K Behera, and Prasanta K Panigrahi. Efficient quantum algorithm for solving travelling salesman problem: An IBM quantum experience. arXiv preprint arXiv:1805.10928, 2018.
25. Abhimanyu Nowbagh and K. Bikash. A quantum approach for solving vehicle routing problem: An IBM quantum experience.
26. Carla Silva, Ines Dutra, and Marcus S Dahlem. Driven tabu search: A quantum inherent optimisation. arXiv preprint arXiv:1808.08429, 2018.
27. Mitali Sisodia, Abhishek Shukla, Alexandre AA de Almeida, Gerhard W. Dueck, and Anirban Pathak. Circuit optimization for IBM processors: A way to get higher fidelity and higher values of nonclassicality witnesses. arXiv preprint arXiv:1812.11602, 2018.
28. Alberto Peruzzo, Jarrod McClean, Peter Shadbolt, Man-Hong Yung, Xiao-Qi Zhou, Peter J Love, Alan Aspuru-Guzik, and Jeremy L O'brien. A variational eigenvalue solver on a photonic quantum processor. Nature Communications, 5(1):1–7, 2014.
29. Nikolaj Moll, Panagiotis Barkoutsos, Lev S Bishop, Jerry M Chow, Andrew Cross, Daniel J Egger, Stefan Filipp, Andreas Fuhrer, Jay M Gambetta, Marc Ganzhorn, et al. Quantum optimization using variational algorithms on near-term quantum devices. Quantum Science and Technology, 3(3):030503, 2018.
30. Lov K Grover. A fast quantum mechanical algorithm for database search. In Proceedings of the Twenty-Eighth Annual ACM Symposium on Theory of Computing, pages 212–219, 1996.
31. Austin Gilliam, Marco Pistoia, and Constantin Gonciulea. Optimizing quantum search using a generalized version of Grover's algorithm. arXiv preprint arXiv:2005.06468, 2020.
32. https://qiskit.org/documentation/tutorials/optimization/index.html.
33. https://www.ibm.com/blogs/research/2020/07/quantum-optim-module/.
34. https://developer.ibm.com/docloud/documentation/optimization-modeling/modeling-for-python/.

4 Multipartite Adaptive Quantum-Inspired Evolutionary Algorithm to Reduce Power Losses of a Radial Distribution Network

4.1 INTRODUCTION

Over the last few decades, some new approximation algorithms have emerged with the aim to explore the search space, which is commonly known as metaheuristics. Generally, metaheuristic is defined as an iterative process which guides the heuristic method by combining the intelligent concepts for exploiting and exploring the search space. If a heuristic optimization algorithm is expressed in a metaheuristic framework with different intelligent concepts to explore the search space is also referred as metaheuristic. Glover introduced the term metaheuristic in 1986 by combing the Greek prefix Meta with heuristic. Heuristic means to find or discover, whereas the suffix Meta means beyond or higher level solution [1]. Metaheuristic method allows the local search operators to escape from local optima by generating new initial solutions or allowing worsening moves for the local search in an intelligent way. High quality solutions are produced in metaheuristics by introducing a bias with various forms [2].

Metaheuristic methods have demonstrated to the scientific community that they are often feasible, alternative, and superior to more traditional methods such as dynamic programming and branch and bound, etc. In comparison with traditional methods, metaheuristics are often providing a better trade-off between computing time and solution quality for large and complicated problems. Metaheuristic methods are often more flexible than traditional methods in two different ways. Firstly, these are adapted to fit for most real-life optimization problems in terms of computational time and solution quality, which can vary greatly across different situations. Secondly, they do not offer any demands on formulation of optimization problem. Metaheuristic methods are implemented by several commercial vendors in their software as primary optimization engine.

Metaheuristic algorithms attempt to find the best feasible solution of an optimization problem out of all possible solutions. Series of operations are performed on the optimization problem to search for the better solution. Local search methods and population based methods are normally employed by metaheuristics to obtain the feasible solutions. Local search methods use iterative process to find the optimal solutions [3]. Population-based methods find the optimal solution by iteratively selecting and then combining existing solutions from a set, usually called as population. The most important member of this class, which mimics the principle of natural evolution, is evolutionary algorithm (EA). In EA, selection operator generally gives direction to the search process by using Darwinian principle. The solution for next iteration is generated through variation operators like crossover by recombining the solutions from current iteration. Local heuristics and mutation operator are used to improve the exploration and exploitation, that is, escaping from local minima and increasing the convergence rate. EAs are popular due to their ease of implementation and employed for solving difficult and complex optimization problems. However, EA often suffers from some limitations.

Quantum-inspired Evolutionary Algorithm (QiEA) is used to overcome these limitation, which use probabilistic representation along with some concepts and operations of quantum computing [4]. It uses a single qubit with small population size and is governed by principle of quantum mechanics [5]. Q-gates are used in QiEA as a variation operator to drive the individuals in the population towards better solution. In recent times, Adaptive Quantum-inspired Evolutionary Algorithm (AQiEA) [6] is applied on various engineering optimization problems with a measurement operator, which is a modified version of QiEA. AQiEA uses two sets of qubits, whereas QiEA uses a single set of qubit. Recently, AQiEA is applied on optimization problem of Distributed Generator (DG) [7]-[8], Network Reconfiguration [9]-[11], ceramic grinding [12], Cost analysis of DG and Capacitor [13], Siting and sizing of Capacitors [14], and simultaneous implementation of both DG and capacitors [15]. In this chapter, we are proposing Multipartite Adaptive Quantum-inspired Evolutionary algorithm , which is an updated version of AQiEA. MAQiEA improves on both exploration and exploitation ability of AQiEA by introducing changes in Rotation towards Better Strategy and Rotation away from Worse Strategy. In MAQiEA, the Rotation toward Better Strategy of AQiEA is converted into Rotation Around Better Strategy as it offers more un-restricted exploration as compared to the previous strategy, which allowed exploration in only improving direction. Similarly, the Rotation away from Worse Strategy in AQiEA involved only two individuals, one was Best Individual and the other was sequential selected Individual and was primarily used for exploitation purpose, i.e., searching around the Best individual with the help of other individuals in the population, so it was bipartite. However, recent algorithms like Grey Wolf Optimizer (GWO) [16], Symbiotic Organism Search [17] and Salp Swarm Algorithm (SSA) [18] etc. are shown to be good exploiters and they tend to use multiple individuals based variation operators rather than bipartite variation operators, therefore, it was decided to augment the exploitation strategy, i.e., Rotation away from worse by converting it into Multipartite Adaptive Variation operator from the current bipartite version in AQiEA.

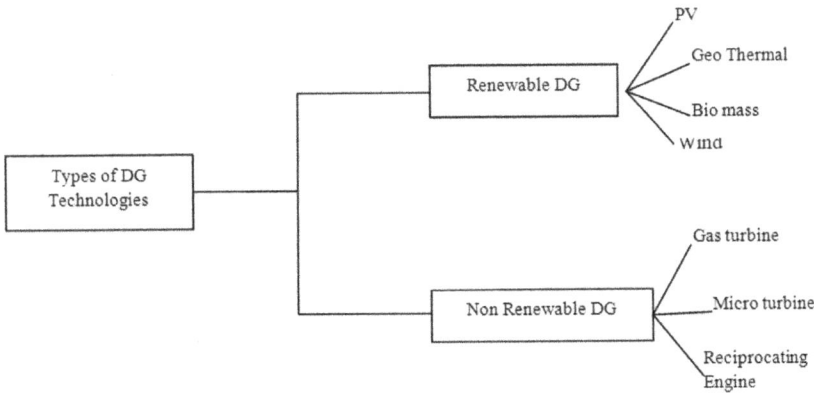

Figure 4.1: Types of DG Technologies.

Distributed Generator: Optimal location and capacity of DG play a key role to minimize the power losses in DN. In general, DG is defined as the small scale power generating source which is placed very nearer to the load centers i.e., supplies electric power near to point of consumption, which reduce the transmission cost and enhance the percentage of power loss reduction in the system [19]. Definition of DG differs from one agency to another [20]. DG is a power generating source used near load centers, which is not connected to high voltage transmission system directly. In comparison with conventional power plants, DG has modular and small size, low installation time with low investment. DGs are operated as power backup supply devices. Implementing the DGs into distribution system has great impact on protection, operation and stability of the system. Figure 4.1 shows the different types of DG technologies, which comprise of Renewable as well as non-renewable energy resources. Figure 4.2 shows the diagrammatic representation of DG capacities. The renewable DG technologies are PV Solar, Geo Thermal, Bio-Mass and Wind turbine, whereas non-renewable technologies are Gas-Turbines, Micro-turbine, reciprocating engines and fuel cells etc. Integrating DG into distribution network have both positive and negative impacts. Positive impact is improvement in percentage power loss reduction, enhancement in voltage profile, improvement in reliability and loadability. Whereas negative impact indicates increment in the short circuit current and reversed power flow. These impacts may vary depending on the optimal placement, sizing and type of DG. Inappropriate location and capacity of DG leads the system to increase in power loss and reduction in voltage profile. Distribution network has dynamic load structure which is not fixed for certain length of time. Industrial, commercial and residential loads are mainly dependent on voltage, which varies with time. However most of the work done in DN to reduce the losses with DG implementation has considered only constant power load (CP) model, which is independent of voltage and doesn't vary with time. In this study, an investigation has been performed with

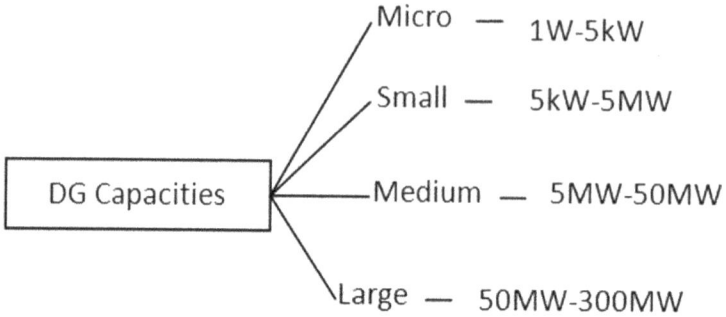

Figure 4.2: DG Capacities.

voltage dependent load model (VDLM) to find the power losses incurred in the system. In this study constant current (CC) and constant impedance (CZ) which varies linearly and square of the voltage are considered. The industrial load (IL), commercial load (CL) and residential load (RL) power requirements vary exponentially with the terminal voltage. In addition to VDLM, CP load model is also used in the study. A class of mixed load (ML) is also investigated to find the power losses incurred in the system i.e., combination of all load including both VDLM and CP load model.

4.2 LITERATURE REVIEW

Research on optimization of DG in DN is becoming more and more popular due to its ease of implementation and ecofriendly technology. Developed and under developing countries are focusing more on renewable energy sources to meet the required load demand with both dispatchable and un-dispatchable DGs. Optimal placement and capacity of DG is one of the most complex optimization problems of electrical power systems. Over the past few decades, optimal location and sizing of DG has become an interesting and challenging area of research to reduce the power losses. Many authors have tried to solve this important optimization problem with three different techniques. In general, these techniques are classified as Analytical, Numerical and Metaheuristic methods. Analytical techniques requires less computational time and easy to implement on smaller bus systems [21]. However, for larger and complex test bus system, its computational complexity is very high. Similarly, numerical methods are difficult to implement on larger test bus systems when new constraints are added [21]. Many techniques have been implemented in DN for optimal location and capacity of DG to reduce the losses. In comparison with analytical techniques, metaheuristic techniques have more ability to provide significant solution. Thus more and more metaheuristic techniques are applied to solve this important optimization problem.

Vijay babu et al. [22] used an analytical approach with real power loss expression to find the optimal location and capacity of DG. An investigation has been performed with four different scenarios to minimize the losses with different types of DGs. All class of mix DGs viz., Type-I (which injects only active power), Type-II (both injects both active and reactive powers), Type-III (only reactive power), is also used in this study as fourth scenario. Prakash and Kathod [23] presented an analytical technique to reduce losses with implementation of single DG. In this study, optimization of DG reduces the magnitude of active and reactive power components. A hybrid technique i.e., combination of both analytical and metaheuristic technique (genetic algorithm) is used in ref [24], to determine the location and capacity of DG. An investigation has been performed on DGs mode of operation, two different scenarios are used with different power factors. Ahmed et al. [25] used a linearized model to estimate the optimal capacity of DG with graph flow and Kalman filter. Optimal size of DG is obtained with a two stage method, where graph flow is used to create a linear model and Kalman filter is used to find the optimal size of DG. Mahmoud et al. [26] presented an analytical approach with integration of DG in distribution system to reduce losses. Multiple DGs are installed in the system with optimal power factors and four different scenarios are used to maximize the percentage power loss reduction with different types of DGs.

Khoa et al. [27] proposed an optimization technique known as one rank cuckoo search algorithm (ORCSA). ORCSA is used to solve the combinatorial optimization problem of DG by finding its optimal location and capacity with different power factors. Multiple objective optimizations are used in the study, Similar to the above, Sultana et al. [16] studied the effect of DG allocation on distribution system with grey wolf optimization algorithm with an objective to reduce the losses. Yifei et al. [28] studied the impact of DG in DN with loss sensitivity factor (LSF), which is used to determine the optimal allocation of DG. Mistry [29] used two different optimization techniques to reduce the losses in a DN with implementation of multiple DGs. Carvalho and Niraldo [30] used an optimization technique known as Ant Colony Optimization for optimization DG with same objective as mentioned above. Vizhiy and Santhi [31] presented a multiobjective optimization problem with DG to reduce the losses. Biogeography based optimization (BBO) is used to find optimal placement and capacity of DG. Ali et al. [32] used evolutionary algorithm technique called as Ant Lion Optimization (ALO), which is used to find the optimal allocation of DG. Proposed algorithm is based on behaviour of hunting lion ants, a multiobjective approach is used to reduce the losses with improvement in voltage profile.

Mahajan and Vadhera [33] used particle swarm optimization technique with an objective to minimize the losses in DN by finding the ideal location of DG with optimal size. An investigation has been performed to find the improvement in voltage profile with multiple weight factors. Snigdha and Panigrahi [34] studied the effect of DG in distribution system with multiobjective differential evolution algorithm to maximize the benefits of DG owners and utilities by minimizing the power losses with different scenarios. Chaotic symbiotic organism search algorithm with multi objective optimization problem is used in ref [35] to minimize the losses in a DN by allocating the DG at optimal location with capacity. Benchmark test bus systems are also used to show the effectiveness of CSOS. Power loss reduction index is considered as primary objective to reduce the losses with implementation of DG, Selective Particle Swarm Optimization (SPSO) determines the optimal placement and capacity of DG [36]. Sarfaraz et al. [37] used same algorithm to find the optimal location and capacity of DG with same objective. Multiple DGs are used with small rating DGs in [36], whereas ref [37] uses a single DG with high operating power (size). Power loss reduction with multiple DGs is high in comparison with single DG.

Devang and Ritesh [38] studied the effect of DG with different power factors i.e, lagging power factor, 0.8 power factor and unity power factor on distribution system at various locations. In this study, two DGs are employed with small size to minimize the losses. Jamian et al. [39] used Gravitational search algorithm to reduce the losses in the DN by placing the DG at its optimal location with optimal capacity. Zhang and Bo [40] studied the impact of DG in radial DN on power loss minimization. Simulation results conclude that, DG should be placed at the end of the line when it is operating with low rating and at the middle of the line when it is operating with high power rating. DG should be placed nearer to substation when it has high operating power greater than the load demand. An optimization technique known as Genetic Algorithm is used to find the placement and capacity of DG with an objective to minimize the power losses [41]. Ang et al. [42] used a new metaheuristic technique viz., sine cosine algorithm to find the optimal placement and capacity of single and multiple DGs in DN with an objective to improve the voltage profile in the network and maximize the percent power loss reduction in the system. Simultaneous implementation of DG and Capacitors are also used by some authors to reduce the power losses. In such studies, DG injects only active power with unity power factor into the system, whereas Capacitor injects only reactive power with zero power factor into the system, injection of both powers into the system results in high reduction in power losses as compared with independent implementation. However, independent implementation of DG has high reduction in power losses in comparison with independent implementation of Capacitors. Simultaneous implementation of both DG and capacitor induces high investment cost, maintenance and operation cost. In our study, only independent implementation of DG is considered.

It has been observed from the above literature that only constant power load is used which doesn't vary with time. In distribution network, consumers use different load models with different ratings. Majority of load used in distribution network is dependent on voltage, however CP load model is independent of voltage. If the optimal placement and capacity of DG obtained with CP load model is used in practical distribution network, which induces more power losses into the system due to the improper location and sizing of DG. Some have used different load models other than CP load. Roy et al. [43] studied the impact of DG in a DN on different load models with voltage profile. Initial investigation has been performed by analyzing the impact of static load (CP load) on DN. Dynamic analysis shows that composite load model has high voltage dips and CZ load model has low voltage dips. Oscar et al. [44] used a novel approach to minimize the power losses by varying the load

Nodes or Buses

Constant Power Load — SMPS, Inverters

Constant Impedance Load — Airport, Some Street Lightning

Constant Current Load — Incandescent Lamps

Industrial Load — Electrical Loads used in Industries

Residential Load — Cooking Equipment, Air Conditioning Units

Commercial Load — Lightning Loads of Malls, Restaurants

Mixed Load

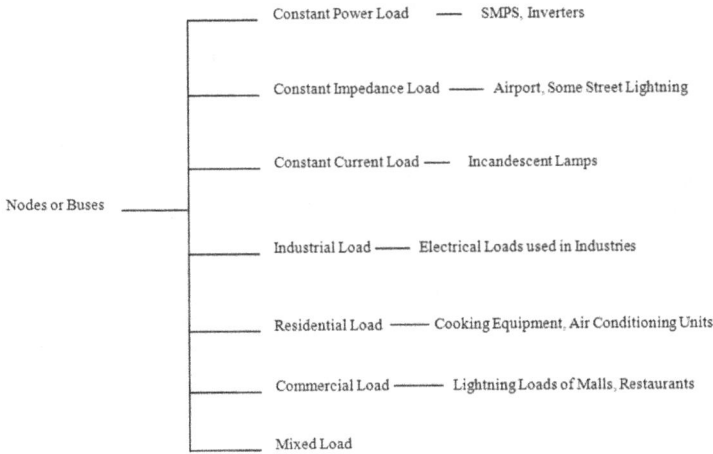

Figure 4.3: Different Load Models.

model i.e., twenty four load model is used for optimization of DG by keeping the voltages within the limits. Divya and Srinivasan [45] studied the effect of DG under fault conditions with a simple radial distribution system. Voltage Stability Index (VSI) is used to find the optimal allocation of DG, whereas its optimal capacity is obtained with Particle Swarm Optimization. Aashish et al. [46] studied the effect of practical load models with integration of DG in DN. Several performance indices are developed as multi-objective function. Genetic algorithm and Particle Swarm optimization are also used to determine the optimal location and size of DG. Das et al. [47] investigated the effect of DG on a VDLM i.e., residential time varying load to reduce the power losses. Sensitivity index-based method is used to find the optimal placement of DG with variation in load whereas, Genetic Algorithm determines the optimal capacity of DG.

In this study, an investigation has been performed to study the effect of DG with variation in load. Different types of load models, which are dependent on exponential characteristics of node voltage, are shown in Figure 4.3. Optimal placement and sizing of DG is a nondifferentiable combinatorial complex optimization problem. Multi-partite Adaptive Quantum-inspired Evolutionary Algorithm (MAQiEA) is used to find the optimal location and capacity of DG for VDLM. MAQiEA is the updated version of AQiEA. In AQiEA, three rotation strategies are used to converge the population towards global optima, whereas MAQiEA uses probabilistic rotation around better and multi-partite rotation away from worse rotation strategies which provides for relatively better exploration and exploitation. MAQiEA has high robustness and has better exploitation and exploration of search space in comparison with AQiEA as shown by test results.

4.3 PROBLEM FORMULATION

DN is normally designed to supply the power to all consumers. But the power demand is increasing day by day due to the vast increase in industrial and human needs. In order to cater the required load demand, distributed generators are used as alternative solution. It is a well-known fact that the load at distribution system changes continuously. In this work, an investigation has been performed to study the effect of DG on VDLM other than CP load. IL, RL, CL, CC and CZ are used in this study [48]-[51]. These are expressed as follows:

$$P_L(m) = P_{L0}(m) \left(S_1 \left(\frac{V_i(m)}{V_o(m)} \right)^{\mu_{cp}} + T_1 \left(\frac{V_i(m)}{V_o(m)} \right)^{\mu_i} + U_1 \left(\frac{V_i(m)}{V_o(m)} \right)^{\mu_r} \right)$$
$$+ P_{L0}(m) \left(V_1 \left(\frac{V_i(m)}{V_o(m)} \right)^{\mu_c} + W_1 \left(\frac{V_i(m)}{V_o(m)} \right)^{\mu_{ci}} + X_1 \left(\frac{V_i(m)}{V_o(m)} \right)^{\mu_{cc}} \right) \quad (4.1)$$

$$Q_L(m) = Q_{L0}(m) \left(S_2 \left(\frac{V_i(m)}{V_o(m)} \right)^{\gamma_{cp}} + T_2 \left(\frac{V_i(m)}{V_o(m)} \right)^{\gamma_i} + U_2 \left(\frac{V_i(m)}{V_o(m)} \right)^{\gamma_r} \right)$$
$$+ Q_{L0}(m) \left(V_2 \left(\frac{V_i(m)}{V_o(m)} \right)^{\gamma_c} + W_2 \left(\frac{V_i(m)}{V_o(m)} \right)^{\gamma_{ci}} + X_2 \left(\frac{V_i(m)}{V_o(m)} \right)^{\gamma_{cc}} \right) \quad (4.2)$$

For Load Type I: Constant Power Load: $S_1 = 1$ and $T_1 = U_1 = V_1 = W_1 = X_1 = 0$
$S_2 = 1$ and $T_2 = U_2 = V_2 = W_2 = X_2 = 0$
For Load Type II: Industrial Load: $T_1 = 1$ and $S_1 = U_1 = V_1 = W_1 = X_1 = 0$
$T_2 = 1$ and $S_2 = U_2 = V_2 = W_2 = X_2 = 0$
For Load Type III: Residential Load: $U_1 = 1$ and $S_1 = T_1 = V_1 = W_1 = X_1 = 0$
$U_2 = 1$ and $S_2 = T_2 = V_2 = W_2 = X_2 = 0$
For Load Type IV: Commercial Load: $V_1 = 1$ and $S_1 = T_1 = U_1 = W_1 = X_1 = 0$
$V_2 = 1$ and $S_2 = T_2 = U_2 = W_2 = X_2 = 0$
For Load Type V: Constant Impedance Load: $W_1 = 1$ and $S_1 = T_1 = U_1 = V_1 = X_1 = 0$
$W_2 = 1$ and $S_2 = T_2 = U_2 = V_2 = X_2 = 0$
For Load Type VI: Constant Current Load: $X_1 = 1$ and $S_1 = T_1 = U_1 = V_1 = W_1 = 0$
$X_2 = 1$ and $S_2 = T_2 = U_2 = V_2 = W_2 = 0$

Where $P_L(m)$ and $Q_L(m)$ are the active and reactive power loads at mth bus. Table 4.1 shows the active and reactive power components of voltage dependent load model [17], [46]. The objective of the study is to reduce the power losses with optimal location and capacity of DG. Power losses are expressed as follows:
$F = \min(P_{loss})$

$$Min(P_{loss}) = \sum_{m=1}^{N_b} I_m^2 * R_m. \quad (4.3)$$

Voltage Stability Index: Distribution network has complex structure, poor voltage regulation and high power losses are observed in the system at the end nodes. In

Table 4.1
Exponent Values of Different Voltage Dependent Loads

Load Type	Exponent	
Constant Power Load	μ_{cp}	γ_{cp}
	0	0
Industrial Load	μ_i	γ_i
	0.18	6.0
Residential Load	μ_r	γ_r
	0.92	4.04
Commercial Load	μ_c	γ_c
	1.51	3.40
Constant Impedance Load	μ_{ci}	γ_{ci}
	1	1
Constant Current Load	μ_{cc}	γ_{cc}
	2	2

order to maintain the system voltage within acceptable limits, voltage stability index is considered.

$$Min(VSI_{k+1}) = V_k^4 - 4(P_{k+1}X_k - Q_{k+1}R_k)^2 - 4(P_{k+1}R_k - Q_{k+1}X_k)V_k^2 \quad (4.4)$$

Power loss minimization in distribution network with optimal placement and sizing of DG is an interesting and challenging area of research. Optimal location and capacity of DG minimizes power losses and improves the voltage profile in the system. Power injected by at a particular bus, m is given as follows.

$$P_L(m) = P_L(m) - P_{DG}(m) \quad (4.5)$$

Constraints for power loss minimization with DG are given as follows:
a) *Operation of Distributed Generator:* DG is used to minimize the power losses with optimal location and capacity, inappropriate sizing of DG induces high power losses in the system. It is necessary that optimal size of DG has to be in specified limits.

$$P_{DG,i}^{min} < P_{DG,i} < P_{DG,i}^{max} \quad (4.6)$$

b) *Power injection for the system:* DGs are placed nearer to load centers, if DG induces high power into the system i.e., greater than total power demand and losses, bi directional power flow may damage the system equipment. It is necessary that, total power injected by DG always less than demand and losses in the system.

$$\sum_{m=1}^{n} P_{DG,i} < P_{Demand} + P_{loss} \quad (4.7)$$

c) Power injection: Total power injected by substation and Distributed generator has to meet power demand at load centers including losses.

$$P_{Substation} + \sum_{m=1}^{n} P_{DG,i} < P_{Demand} + P_{loss} \tag{4.8}$$

d) Voltage limit: Optimal location and capacity of DG not only reduces the power losses but also improves the voltage profile in the system. After installing DG at optimal location with capacity, voltage has to be in permissible limits.

$$V^{min} < V < V^{max} \tag{4.9}$$

4.4   POWER FLOW

In a power system network, ac and dc power flows from generating station to the utilities at load end through different branches and busses. The flow of power i.e., active or reactive in a system is called load flow or power flow. Load flow or power flow study is generally a mathematical approach used to determine the load under steady state condition, active or reactive power flow through different branches, generators, total active and reactive power losses, individual power losses obtained at each branch and bus voltages. Load flow analysis is used to determine the steady state operation of network. It is widely used by power system professional during planning stages of distribution system or adding an additional network to existing one. Traditional power flow methods such as Gauss-Seidel, fast decoupled and Newton Raphson methods are used to find the power losses, voltage profile of the network. These methods are mainly designed and used for transmission systems and unsuitable for distribution networks. Distribution system has complex structure with large number of branches and nodes. Distribution networks are normally designed as mesh structures but in normal operating conditions they are operated in radial structure. The load at distribution network generally poses unbalanced operation with unbalanced distribution loads. Traditional power flow methods require set of equations whose size is equal to number of buses. If the traditional power flow methods are used in distribution network, for large test bus systems the computational time is too long. In addition, Y-bus formation in traditional power flow methods creates waste of memory storage. Hence, traditional power flow methods are inefficient for distribution networks. Distribution system has high R/X ratio, which causes ill conditioning for traditional power flow methods. Hence, there is need for a load flow which uses topological characteristics of distribution network to determine the total power losses and bus voltage in the network. Jen-Hao Teng [52] has proposed a load flow which uses topological characteristic of distribution network. Based on the topological characteristics of the network, Bus Injection to Branch Current (BIBC) matrix and Branch Current to Bus Voltage (BCBV) matrix are normally developed. This method generally has three important steps which are given below.

Equivalent current injection
Formation of BIBC Matrix
Formation of BCBC Matrix

Equivalent Current Injection: Equivalent current injection of the distribution system at bus k, the complex power S_k for t^{th} iteration is given as

$$S_k = P_k + jQ_k \qquad (4.10)$$

$$I_k^t = I_k^r(V_k^t) + jI_k^i(V_k^t) \qquad (4.11)$$

Formation of BIBC Matrix: By applying Kirchhoffs Current Law (KCL) to the simple radial distribution network shown in Figure 4.4, the equivalent current injection obtained is

$$B_2 = I_3 + I_4 B_4 = I_5 + I_6 + I_7 B_5 = I_6 + I_7 + I_8 B_7 = I_8 \qquad (4.12)$$

Bus Injection to Branch Current (BIBC) matrix is obtained as follows

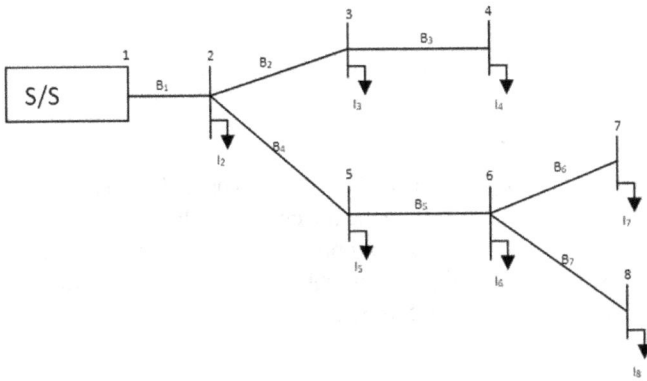

Figure 4.4: Radial Distribution System.

$$
\begin{bmatrix} B_1 \\ B_2 \\ B_3 \\ B_4 \\ B_5 \\ B_6 \\ B_7 \end{bmatrix}
=
\begin{bmatrix}
1 & 1 & 1 & 1 & 1 & 1 & 1 \\
0 & 1 & 1 & 0 & 0 & 0 & 0 \\
0 & 0 & 1 & 0 & 0 & 0 & 0 \\
0 & 0 & 0 & 1 & 1 & 1 & 1 \\
0 & 0 & 0 & 0 & 1 & 1 & 1 \\
0 & 0 & 0 & 0 & 0 & 1 & 0 \\
0 & 0 & 0 & 0 & 0 & 0 & 1
\end{bmatrix}
\begin{bmatrix} I_2 \\ I_3 \\ I_4 \\ I_5 \\ I_6 \\ I_7 \\ I_8 \end{bmatrix}
$$

The relation between equivalent current injection and branch currents are expressed as

$$[B] = [BIBC][I] \qquad (4.13)$$

Algorithm for formulation of BIBC Matrix:
Step I: If a distribution network has u branch section and v bus section, a null matrix with dimension u*(v-1) is created.

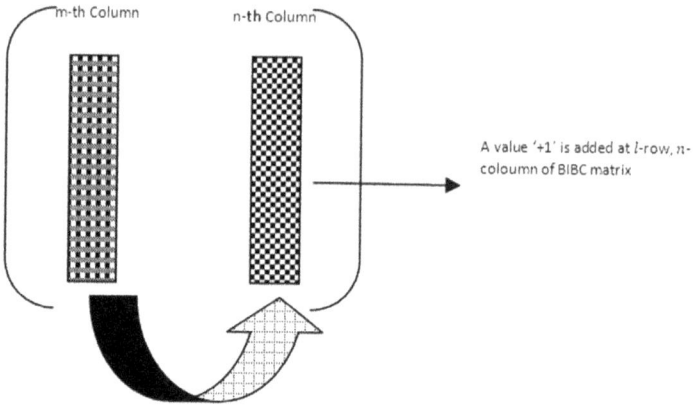

Figure 4.5: Diagrammatic representation of BIBC matrix.

Step II: If a line section (B_L) is located between two buses bus m and bus n, copy the column of mth bus of BIBC matrix to the nth column of BIBC matrix and a value +1 is added at a position of l-row, n-bus column. If a line section (B_5) is located between two buses bus 6 and bus 8, copy the 6th column of BIBC matrix to 8th column of BIBC matrix and +1 is added at a 5-row, 8-column of BIBC matrix.

Step III: The above process is repeated for all branch line sections in the distribution network.

The above process of copying the columns from one column to other column and addition of value +1 is shown in Figure 4.5.

Formation of BCBV Matrix:

Branch-current and bus voltage relationship (BCBV) is obtained by applying KVL to the simple radial distribution network shown in Figure 4.6.

$$V_2 = V_1 - B_1 Z_{12} V_5 = V_2 - B_4 Z_{25} V_6 = V_5 - B_5 Z_{56} \qquad (4.14)$$

Using above equation the bus voltage at 6th bus is written as

$$V_6 = V_1 - B_1 Z_{12} - B_4 Z_{25} - B_5 Z_{56} \qquad (4.15)$$

From the above equation, it has been observed that the bus voltage can be expressed as function of substation voltage, branch current and line parameters. Branch Current to Bus Voltage (BCBV) matrix is obtained as follows.

$$
\begin{pmatrix} V_1 \\ V_1 \\ V_1 \\ V_1 \\ V_1 \\ V_1 \\ V_1 \end{pmatrix} - \begin{pmatrix} V_2 \\ V_3 \\ V_4 \\ V_5 \\ V_6 \\ V_7 \\ V_8 \end{pmatrix} = \begin{pmatrix} Z_{12} & 0 & 0 & 0 & 0 & 0 & 0 \\ Z_{12} & Z_{23} & 0 & 0 & 0 & 0 & 0 \\ Z_{12} & Z_{23} & Z_{34} & 0 & 0 & 0 & 0 \\ Z_{12} & 0 & 0 & Z_{25} & 0 & 0 & 0 \\ Z_{12} & 0 & 0 & Z_{25} & Z_{56} & 0 & 0 \\ Z_{12} & 0 & 0 & Z_{25} & Z_{56} & Z_{67} & 0 \\ Z_{12} & 0 & 0 & Z_{25} & Z_{56} & 0 & Z_{68} \end{pmatrix} \begin{pmatrix} B_1 \\ B_2 \\ B_3 \\ B_4 \\ B_5 \\ B_6 \\ B_7 \end{pmatrix}
$$

Therefore branch currents and bus voltages are expressed as

$$(\Delta V) = (BCBV)(B) \tag{4.16}$$

Algorithm for formulation of BCBV Matrix:
Step I: If a distribution network has u branch section and v bus section, a null matrix with dimension (v-1)*u is created.
Step II: If a line section (B_L) is located between two buses bus m and bus n, copy the row of mth bus of BCBV matrix to the nth row of BCBV matrix and fill the line section impedance (Z_{mn}) to the position of the n-th row to l-th column. If a line section (B_5) is located between two buses bus 6 and bus 8, copy the 6th row of BCBV matrix to 8th row of BCBV matrix and an impedance of value Z_{68} is added at a 8-row, 6-column of BCBV matrix.
Step III: The above process is repeated for all branch line sections in the distribution network.
The above process of copying the rows from one row to other row and addition of value $'Z'_{mn}$ is shown in Figure 4.6. *Algorithm for Distribution Load Flow:* Distri-

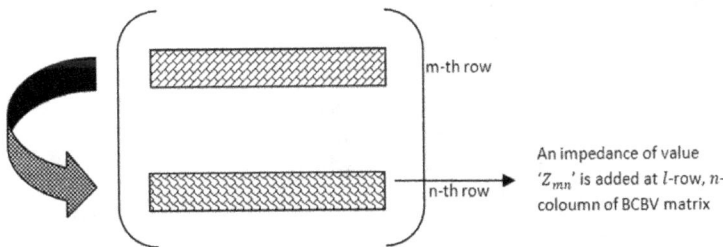

m-th row

An impedance of value
$'Z_{mn}'$ is added at l-row, n-coloumn of BCBV matrix

n-th row

Figure 4.6: Diagrammatic representation of BCBV matrix.

bution load flow is used to find the total power losses induced in the system, bus voltages and individual power losses for each branch. Step-by-step process of direct load flow is given as follows.
Step I: Read the initial line data and load data of the distribution network.

*Step II:*Bus Injection to Branch Current matrix is formulated based on data.

*Step III:*Branch Current to Bus Voltage matrix is formulated based on data.

*Step IV:*Direct load flow matrix is created. The relationship between bus current injection and bus voltage is given by

$\Delta V = [BCBV][BIBC][I]$

$= [DLF][I]$

*Step V:*Initialize the iteration count t=0

*Step VI:*Increment in iteration count t=t+1

*Step VII:*Update the voltage by solving the equation iteratively. By solving the below equations iteratively, the solution of the direct load flow is obtained.

$I_k^t = I_k^r(V_k^t) + jI_k^i(V_k^t)$

$[\Delta V(t+1)] = [DLF][I^t]$

$[V(t+1)] = [V^0][\Delta V(t+1)]$

If tolerance exists or converges print results else go to step (6).

Flow chart is representation of direct load flow is shown in Figure 4.7.

4.5 ALGORITHM

In recent times, metaheuristics are mainly used to solve several optimization problems. Metaheuristic methods have demonstrated the scientific community that they are often feasible, alternative, and superior to more traditional methods such as dynamic programming and branch and bound, etc. In comparison with traditional methods, metaheuristics are often providing a better trade-off between computing time and solution quality for large and complicated problems. In EA, individuals will compete with one another. The fittest individual in the population will move forward to the next generation. It acts as a parent and it will again compete with child in the next generation, the fittest among them will move forward to further generation. This process repeats until convergence exists. EA often suffers from some major limitations i.e., stagnation, sensitivity to the choice of parameter, premature and Slow convergence. QiEA overcomes the above limitations by creating a good balance between exploration and exploitation. QiEA is designed by integrating principles of Quantum mechanics viz., measurement, entanglement, superposition and interference into current framework of EA. It is proposed to solve difficult combinatorial and non-differentiable optimization problems. In this study, A Multi-partite Adaptive Quantum-inspired Evolutionary Algorithm (MAQiEA) is used to solve non linear large-scale optimization problem. AQiEA [9] is different from QiEA, AQiEA uses two Q-bits per solution vector, whereas Quantum-inspired Evolutionary Algorithm uses a single Q-bit. In Adaptive Quantum-inspired Evolutionary Algorithm, the smallest information element in a quantum computer is a quantum-bit (qubit) analogous to classical bits. The basis states are represented in Hilbert space by a vector as $|0\rangle$ and $|1\rangle$. The qubit can be represented by vector $|C\rangle$ and it is defined as

$$|C\rangle = A|1\rangle + B|0\rangle \tag{4.17}$$

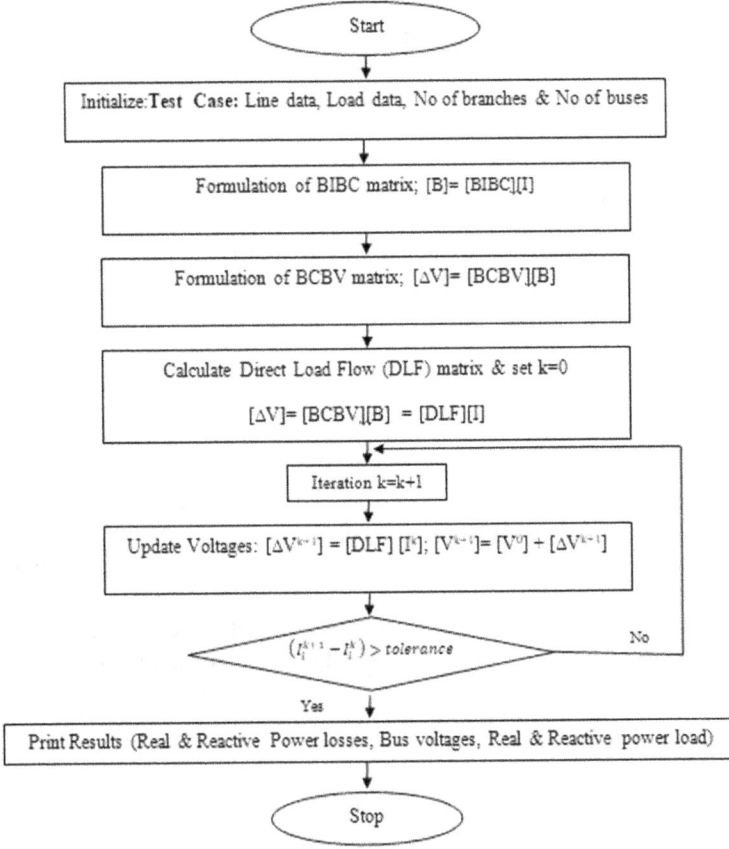

Figure 4.7: Flow Chart of Direct Load Flow.

Where A and B are complex numbers which specifies the probability amplitudes associated with states $|1\rangle$ and $|0\rangle$ respectively and should satisfy the condition.

$$|A|^2 + |B|^2 = 1 \tag{4.18}$$

Where $|A|^2$ and $|B|^2$ specify the probability of qubit to be in state 0 and 1.

The proposed Multi-partite Adaptive Quantum-inspired Evolutionary Algorithm employs two qubits, first qubit is used to store the solution vector of design variables and the second qubit is used to store the scaled and ranked objective function value [8]. The classical implementation of Entanglement principles are mathematically represented as follows.

$$|C_{2i}(t)\rangle = f_1|C_{1i}(t)\rangle \tag{4.19}$$

$$|C_{1i}(t+1)\rangle = f_2(|C_{2i}(t)\rangle, |C_{1i}(t)\rangle, |C_{1j}(t)\rangle) \tag{4.20}$$

Where $|C_{2i}\rangle$ is i^{th} vector of second qubit, $|C_{1i}\rangle$ is i^{th} solution vector of first qubit and $|C_{1j}\rangle$ is j^{th} solution vector of first qubit, t is iteration number, f_1 and f_2 are the functions through which both the qubits are classically entangled. The second qubit is used as feedback in parameter/ tuning free adaptive quantum-inspired rotation crossover operator. A_{1i} is the probability amplitude of the scaled value of ith variable in the ith qubit. The variables are scaled between upper and lower limits, the limits are taken as zero and one. The qubits are stored in quantum register. Number of variables is equal to number of qubits per quantum register Q_i. The structure of Q_i is shown below:

$$Q_{1,i} = [A_{1,i,1}, A_{1,i,2}.....A_{1,i,n}]$$

$$................$$

$$Q_{1,m} = [A_{1,m,1}, A_{1,m,2}, ...A_{1,m,n}]$$

The second set of qubit in quantum register Q_{i+1} is used to store the scaled and ranked objective function value of corresponding solution vector in Q_i. The fittest vector for objective function value is assigned 1, whereas the worst vector for objective function value is assigned 0 of second qubit set. The remaining solution vectors for objective function value of second qubit is also ranked in the range of zero and one.

If nv represents the number of variables used to solve the optimization problem of DG for optimal placement and capacity of DG with total population np.

$$\begin{bmatrix} Q_1 \\ \vdots \\ \vdots \\ Q_{np} \end{bmatrix} = \begin{bmatrix} X_{11} & \cdots & \cdots & Y_{1nv} \\ \vdots & \vdots & \vdots & \vdots \\ \vdots & \vdots & \vdots & \vdots \\ X_{np1} & \cdots & \cdots & Y_{npnv} \end{bmatrix} \qquad (4.21)$$

Minimization of power losses with optimal location and sizing of DG is considered as main objective. The solution vector for solving the above mentioned objective is represented in Figure 4.8 as follows: The solution vector for simultaneous placement

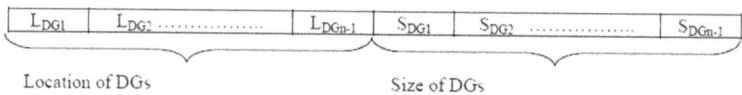

| L_{DG1} | L_{DG2} .... | L_{DGn-1} | S_{DG1} | S_{DG2} .... | S_{DGn-1} |

Location of DGs   Size of DGs

Figure 4.8: Solution vector representation for DG with optimal location and sizes.

and sizing of DG to reduce the power losses is given as follows.

$Q =$

$$\begin{bmatrix} L_{DG1,1} & L_{DG1,2} & \cdots & L_{DG1,np} & S_{DG1,1} & S_{DG1,2} & \cdots & S_{DG1,np} \\ L_{DG2,1} & L_{DG2,2} & \cdots & L_{DG2,np} & S_{DG2,1} & S_{DG2,2} & \cdots & S_{DG2,np} \\ \vdots & \cdots & \cdots & \vdots & \vdots & \cdots & \cdots & \vdots \\ L_{DGnp-1,1} & L_{DGnp-1,2} & \cdots & L_{DGnp-1,np} & S_{DGnp-1,1} & S_{DGnp-1,2} & \cdots & S_{DGnp-1,np} \\ L_{DGnp,1} & L_{DGnp,2} & \cdots & L_{DGnp,np} & S_{DGnp,1} & S_{DGnp,2} & \cdots & S_{DGnp,np} \end{bmatrix}$$

$$(4.22)$$

Three rotation strategies have been applied to converge the population adaptively towards global optima.

Rotation towards the Best Strategy (R-I): All the solution vectors in the population are rotated towards the best solution vector. It is expected that better candidate solution will be found for all other vectors by rotating the remaining solution vectors towards the best solution vector.

Rotation Around the Better Strategy (R-IIA): This strategy is primarily used for exploration purpose, that is two individuals are randomly selected and the search takes place around the better individual. The direction and the magnitude of search region is determined by the relative fitness represented in second set of qubits and relative position of the two individuals stored in the first set of qubits.

Multi Parent Rotation away from worse (R-IIIM): This is inspired from multi parenting strategy, which has been previously used by some metaheuristics such as Grey Wolf Optimization [16] and Symbiotic Organism Search [17]. These metaheuristics are known for their exploitation, so R-IIIM now employs the best individual, a sequentially selected individual, a randomly selected individual and the worst individual in the population.

Flow chart of proposed algorithm with direct load flow is shown in Figure 4.9.

Pseudo code of the Multipartite Adaptive Quantum-inspired Evolutionary Algorithm is shown below:

Pseudo Code:

Initialization

N_p=Number of Quantum Registers i.e., Quantum-inspired Registers Q_1

for i=1: N_p {

Q_1 (i) =rand (0, 1) ;}

Do {

Measurement Operator

for i=1: N_p {

if rand(0,1)¡$(Q_1(i))^2$

$Q_m(i)$= $(Q_1(i))^2$

else

$Q_m(i)$= $(1-(Q_1(i))^2)$}

Fitness calculation

for i=1: N_p {

var(i) = Back transform($Q_m(i)$)

fitness function(i) = DFL PF(var(i))}

Assign Q_2 using fitness level of solution vector of Q_1

Apply Adaptive Quantum based crossover operator using Q_1 and Q_2 to generate Q_{1c}

Elitist selection between Q_{1c} and Q_1

} While (!termination criteria)

Description:

1. Population size, number of variables and maximum number of iterations are initially assigned for quantum register i.e., the Quantum-inspired register Q1 is initialized randomly.

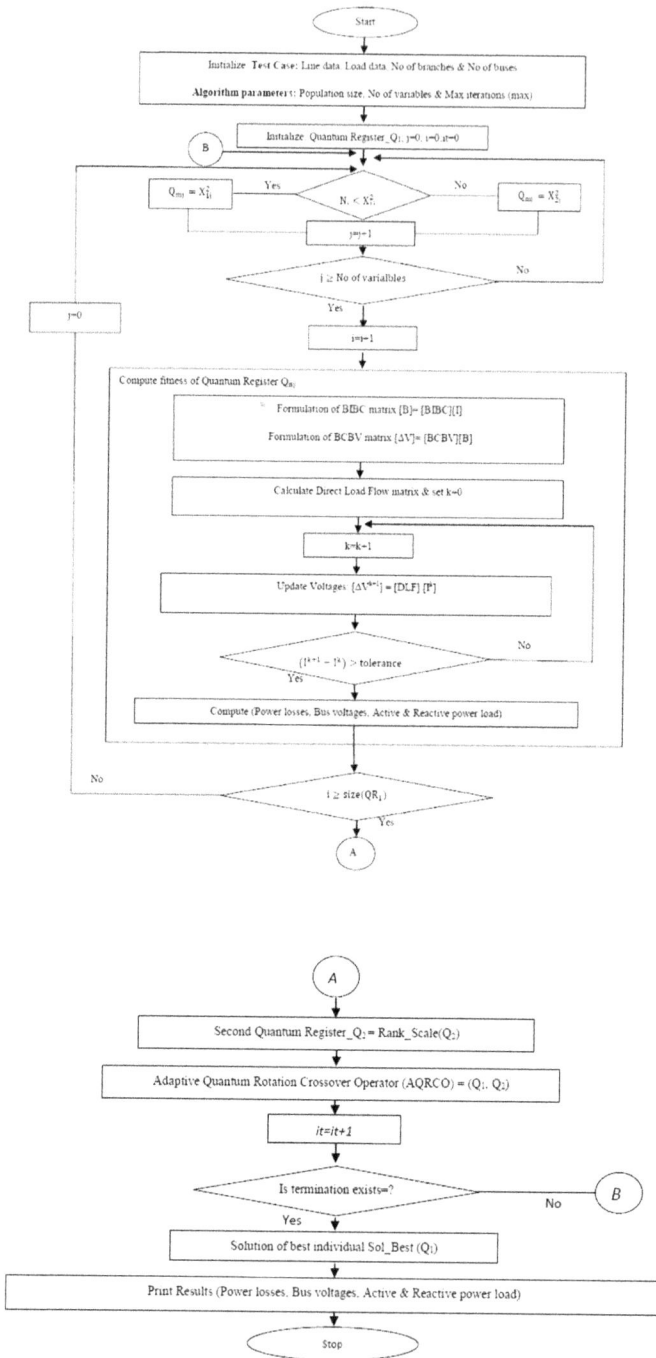

Figure 4.9: Flow chart for the proposed algorithm with Load Flow.

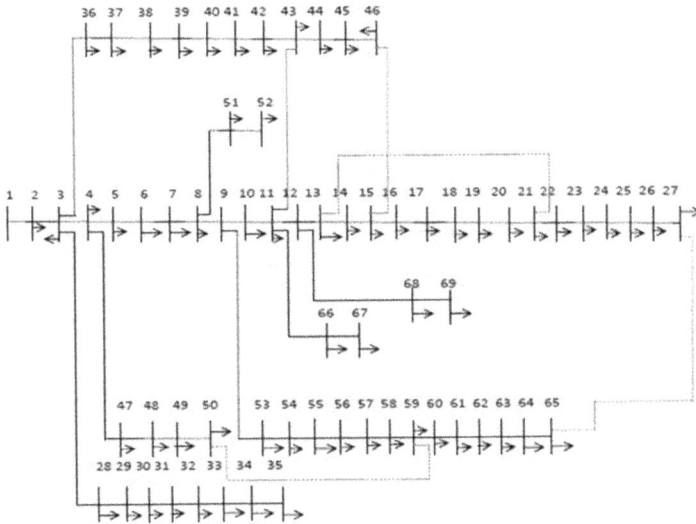

Figure 4.10: Single line diagram of test bus system.

2. A new measured value string (Q_m) is generated by implementing measurement operation on the qubit string.

3. Quantum register Q_m is then back transformed into solution vector, var, i.e. phenotype and is used to compute the fitness through the load flow by Jen-Hao Teng [52] explained in Section Power Flow.

4. Second qubit QR_2 stores scaled and ranked for solution vector within value [0, 1]. The worst solution vector and fittest solution vector in the second qubit is given the value 0 and 1. The remaining solution vectors in the second qubit are arranged between 0 to 1.

5. Three rotation strategies R-I, R-IIA, R-IIIM are applied on Q_1 in Adaptive Quantum Crossover variation operator using Q_2 to generate children population Q_{1c}.

6. By applying tournament selection between Q_1 and Q_{1c}, the corresponding winning individuals amongst Q_1 and Q_{1c} will move to the next generation.

7. Maximum number of iterations is used as the termination criterion.

4.6 RESULTS AND DISCUSSION

The performance of the MAQiEA is validated on IEEE standard benchmark test bus system i.e., 69-bus system to reduce the power losses in distribution network. Under planning stage, distribution network is designed as meshed or looped structure. In normal operating conditions the network is operated in radial structure. Figure 4.10 shows the single line diagram of test bus system. All the buses in the benchmark test

Table 4.2

Initial Load Data of the Test Bus System

Particular Value	(69 Bus System)
Total Active Power Demand (MW)	3.81
Total reactive Power Demand (MVAr)	2.694
Buses	1 - 69
Sectionalizing switches	1 - 68
Tie line switches	69 - 73
Maximum power rating of DG (MW)	1

bus system are considered as candidate nodes for optimal location and capacity of multiple DGs except the substation node. All the parameters in the system i.e., line data and load data of the benchmark test bus system is converted into per units (p.u) for calculation purpose. Experimental results on test bus systems are carried out on MATLAB environment hosted on a Intel Core TM i3 CPU computing machine with 4 GB RAM capacity @1.80GHz. MAQiEA has shown better performance as compared with other algorithms available in the literature. The initial data of the medium voltage test bus system viz., line data and load data are shown in Table 4.2, the total real and reactive power demand of the system i.e., active and reactive loads on the network is 3801kW and 2694kVAr. The 69 bus system mainly consists of sixty nine buses, which has 5 tie line switches and 68 sectionalizing switches which are numbered as 1 to 68 and 69 to 73 respectively. Active and reactive power losses of constant power load model with normally open switches 69, 70, 71, 72 and 73 i.e., without opening any tie line switches and without DG implementation is 224.94 kW and 102.12 kVAr with minimum voltage 0.9092p.u respectively. The parameters used for testing of MAQiEA is given in Table 4.3. First of all, testing has been performed to validate the changes made in the original AQiEA [8] to arrive at MAQiEA. Thereby, testing was done for validating design decision for changing Rotation Towards Better Strategy in AQiEA [7] to Rotation Around Better Strategy (TR-II). Further, testing was also performed for validating the design decision for changing Bipartite Rotation away from Worse Strategy in AQiEA [8] to Multipartite Rotation away from Worse Strategy in MAQiEA. Constant power load model is considered to validate the results.

Tests for validating design decision for Rotation Around Better Strategy (R-IIA):

The R-II in AQiEA was generating a random variable between [0, 1] as is done in majority of EAs and so it was termed as a Rotation towards Better strategy. However, In R-IIA, a random variable is generated between [-1, 1], and is termed as Rotation Around Better Strategy in MAQiEA. We have arrived at [-1, 1], after thorough investigations, which have been performed to find the power losses incurred in the system with different search spaces other than 0 and 1. Seven different search intervals have been investigated to arrive at the best performing one. In Case I, the

Table 4.3

Parameters Different State-of-Art-techniques

GA	Population Size=50; Number of Generations=100; Mutation probability= 0.02; Crossover probability= 0.8
PSO	Population Size=50; Acceleration factor $C_1=C_2=2$; Inertia weights $W_{max}=0.9$, $W_{min}=0.4$
ALO	Number of Agents $N=50$; $Iter_{max}=200$
GSA	Number of Agents $N=50$; $Iter_{max}=200$
MAQiEA	Population Size=50; $Iter_{max}=200$

Table 4.4

Wilcoxon Signed Rank Test on MAQiEA with Different Cases

Comparison	R^+	R^-	P_{Value}	Hypothesis
Case-I Vs Case-II	0	465	0	H_0 : Case-I $\geq$ Case-II H_1 : Case-I $\leq$ Case-II
Case-I Vs Case-III	26	439	0	H_0 : Case-I $\geq$ Case-III H_1 : Case-I $\leq$ Case-III
Case-I Vs Case-IV	26	439	0	H_0 : Case-I $\geq$ Case-IV H_1 : Case-I $\leq$ Case-IV
Case-I Vs Case-V	0	465	0	H_0 : Case-I $\geq$ Case-V H_1 : Case-I $\leq$ Case-V
Case-I Vs Case-VI	71	394	0.0004	H_0 : Case-I $\geq$ Case-VI H_1 : Case-I $\leq$ Case-VI
Case-I Vs Case-VII	0	465	07	H_0 : Case-I $\geq$ Case-VII H_1 : Case-I $\leq$ Case-VII

search interval limits are [-1, 1], for Case II the limits are [0, 1], similarly for other remaining the search intervals as Case III to Case VII are [-0.5, 1], [-1, 0.5], [-0.25, 1], [-0.25, 0.25] and [0.25, 0.5], respectively. The proposed algorithm is tested with different search intervals to minimize the fitness function. It was observed [-1 1], has minimum fitness value in comparison with others. Wilcoxon signed rank test is performed between Case I and other Cases to test the statistical significance of Cass I. In Wilcoxon signed rank test two hypothesis are created, Null Hypothesis and Alternate Hypothesis i.e., H_0 and H_1, respectively and significance level, α is 0.05, which is compared with the P_{Value} to arrive at a conclusion [53]-[54] shown in Table 4.4. Table 4.4 shows, the pair wise comparison of Case-I with other Cases. Case-I shows significant improvement over other Cases with level of significance α =0.05. It has been observed from tabulated results that null hypothesis is rejected based on P_{Value} i.e., level of significance, hence Case I is the best performing search interval.

Table 4.5

Performance Analysis of MAQiEA with Other Alternatives

	AQiEA	$MAQiEA_1$	$MAQiEA_2$	MAQiEA
St. Dev	0.6573	0.0908	0.0984	0.0821
Average	73.4776	71.9906	71.9624	71.8834

Tests for validating design decision for Multipartite Rotation away from Worse Strategy: In MAQiEA, a novel rotation strategy i.e., Multi Parent Rotation away from worse is used, R-IIIM, which now employs the best individual, a sequentially selected individual, a randomly selected individual and the worst individual in the population. Whereas R-III, in AQiEA, had used the best individual, and a sequentially selected individual. In order to show the effectiveness of MAQiEA, two algorithms other than AQiEA, is incorporated in the study viz., $MAQiEA_1$ and $MAQiEA_2$. $MAQiEA_1$ uses tripartite multi parent R-III strategy with best individual, a sequentially selected individual, and a randomly selected individual from the population, whereas $MAQiEA_2$ uses tripartite multi parent strategy with best individual, a sequentially selected individual, and the worst individual in the population. Each algorithm is analysed with thirty independent runs. Based on the performance of theses runs minimum power loss, maximum power loss, average power loss and standard deviation are calculated. In addition, Wilcoxon signed rank test is also used to validate the results statistically. Null hypothesis and alternate hypothesis are represented as H_0 and H_1 for each test case. Table 4.5 shows the power losses i.e., minimum, maximum and average power losses after thirty independent runs. Table 4.6 shows, the pair wise comparison of MAQiEA and other algorithms. It has been observed from tabulated results that null hypothesis is rejected based on P_{Value} i.e., level of significance. That is MAQiEA is the best performing version on the basis of Average and Standard deviation as well as on the basis of Wilcoxon Signed Rank test. For constant impedance load model, total active and reactive power losses induced in the system before installing DG is 188.6958kW, 86.5806kVAr with minimum 0.9174p.u. Similarly, for constant current load model, real and reactive power losses obtained for base case i.e., without implementation of DG is 158.878kW and 73.767kVAr with minimum VSI 0.7305p.u. For industrial load model, the load used in the test bus system is totally dependent on industrial load. Similarly for residential and commercial load models also load used in the system is purely dependent on residential and commercial loads. In case of constant current load and constant impedance load model the total load varies with square of the voltage and linearly with voltage respectively. Power losses obtained for practical load models without implementing DG for industrial, residential and commercial load is171.4316kW, 164.9382kW and 157.0083kW, respectively. A class of mix load is also considered in the study, which incorporates all voltage-dependent load models including constant power load which is independent of voltage. The total active and reactive power losses obtained before implementing DG is 208.8178kW and 95.0525kVAr respectively.

Table 4.6
Wilcoxon Signed Rank Test on MAQiEA with Other Algorithms

Comparison	R^+	R^-	P_{Value}	Hypothesis
AQiEA Vs $MAQiEA_1$	464	1	0	$H_0 : \text{AQiEA} \leq MAQiEA_1$
				$H_1 : \text{AQiEA} \geq MAQiEA_1$
AQiEA Vs $MAQiEA_2$	464	1	0	$H_0 : \text{AQiEA} \leq MAQiEA_2$
				$H_1 : \text{AQiEA} \geq MAQiEA_2$
AQiEA Vs MAQiEA	464	1	0	$H_0 : \text{AQiEA} \leq \text{MAQiEA}$
				$H_1 : \text{AQiEA} \geq \text{MAQiEA}$
$MAQiEA_1$ **Vs MAQiEA**	438	27	0	$H_0 : MAQiEA_1 \leq \text{MAQiEA}$
				$H_1 : MAQiEA_1 \geq \text{MAQiEA}$
$MAQiEA_2$ **Vs MAQiEA**	366	99	0.03	$H_0 : MAQiEA_2 \leq \text{MAQiEA}$
				$H_1 : MAQiEA_2 \geq \text{MAQiEA}$

The parameters used in the algorithms are given in Table 4.3. For constant power load model, tabulated results in Tables 4.7, 4.8 and 4.9 demonstrate that, MAQiEA has high reduction in power losses of 71.7497kW and 35.9981kVAr with minimum voltage 0.979 by implementing DG at optimal location 60, 61 and 17 with optimal sizes 1MW, 774kW and 509kW respectively. Whereas, GA has minimum reduction in power losses as compared with all other algorithms. GA has power loss reduction of 89.737kW and 43.2099kVAr with minimum voltage 0.9093 at optimal location 61, 12, and 23 with capacities 1MW, 520kW and 500kW respectively. ALO has power loss reduction of 78.1456kW and 38.5394kVAr with optimal location 68, 62 and 63 and capacity of 361kW, 933kW and 935kW. Placing DG at optimal location with optimal capacity not only reduces the power loss but also improves the voltage profile of the system. MAQiEA has better improvement in voltage profile in comparison with GA, PSO, GSA and ALO.

For constant impedance load, proposed algorithm has high reduction in power loss of 61.3645kW and 31.4816kVAr with implementation of DG at 61, 16 and 60 with capacities of 639kW, 486kW and 1MW respectively. Except the proposed algorithm, ALO has high power loss reduction of 65.3053kW and 32.8232kVAr with location 61, 68 and 62 and sizing of 1MW, 410kW and 750kW in comparison with other algorithms. MAQiEA has high improvement in VSI in comparison with GA, GSA, PSO and ALO. GA has better improvement in voltage profile and VSI in comparison with PSO and GSA, however in this case also GA minimum loss reduction in power loss of 77.8474kW and 38.6822kVAr with DG location 65,55 and 61 and with capacity of 976kW, 940kW and 467kW respectively.

Similarly for constant current load, ALO has high reduction in power loss except MAQiEA. MAQiEA has maximum power loss reduction in comparison with other algorithms. The overall active and reactive power loss obtained after implementing DG with ALO is 59.1839kW and 30.0164kVAr at locations 64, 60 and 68 with optimal capacities 736kW, 691kW and 900kW respectively. Whereas, PSO and

Table 4.7

Comparative Analysis of MAQiEA with Other Algorithms.

		Base Case	GA	PSO	GSA	ALO	MAQiEA
Constant Power Load	Location	...	61, 12, 23	68, 57, 61	60, 63, 56	68, 62, 63	60, 61, 17
	Size (MW)	...	1, 0.52, 0.5	0.87, 1, 0.88	0.7964, 0.9312, 0.9657	0.3607, 0.9326, 0.9351	1, 0.7733, 0.5084
	$P_{loss}(kW)$	224.9244	89.737	83.2238	84.2603	78.1456	71.7497
	$Q_{loss}(kW)$	102.1108	43.2099	39.8034	41.1438	38.5394	35.9981
	$V_{min}(p.u)$	0.9093	0.9541	0.9714	0.9731	0.9745	0.979
	VSI(p.u)	0.6827	0.8277	0.8895	0.8968	0.9016	0.9177
	$P_{load}(MW)$	3.8019	1.7819	1.0519	1.1086	1.5735	1.5202
	Location	...	65, 55, 61	68, 11, 61	15, 60, 63	61, 68, 62	61, 16, 60
	Size (MW)	...	0.9751, 0.9394, 0.4667	0.4766, 0.6815, 1	0.7108, 0.6714, 0.5972	1, 0.4109, 0.7493	0.6386, 0.4855, 1
Constant Impedance Load	$P_{loss}(kW)$	188.6958	77.8474	73.3395	68.6522	65.3053	61.3645
	$Q_{loss}(kW)$	86.5806	38.6822	35.9269	34.5953	32.8232	31.4816
	$V_{min}(p.u)$	0.9174	0.9736	0.9624	0.9682	0.977	0.9813
	VSI(p.u)	0.7076	0.8986	0.8569	0.8765	0.9109	0.9263
	$P_{load}(MW)$	3.619	1.2378	1.4609	1.6316	1.4588	1.4584
	Location	...	27, 64, 49	61, 50, 68	60, 55, 13	64, 60, 68	16, 60, 63
	Size (MW)	...	0.7165, 0.9724, 0.7993	1, 0.3386, 0.4249	0.9561, 0.8975, 0.1449	0.7356, 0.6909, 0.9003	1, 0.5324, 0.4384

Multipartite Adaptive QIEA to Reduce Power Losses

Table 4.8

Comparative Analysis of MAQiEA with Other Algorithms (Contd...)

		Base Case	GA	PSO	GSA	ALO	MAQiEA
Constant Current Load	Location	...	61, 12, 23	68, 57, 61	60, 63, 56	68, 62, 63	60, 61, 17
	Size (MW)	...	1, 0.52, 0.5	0.87, 1, 0.88	0.7964, 0.9312, 0.9657	0.3607, 0.9326, 0.9351	1, 0.7733, 0.5084
	$P_{loss}(kW)$	158.8787	71.1685	63.5002	66.9969	59.1839	52.6348
	$Q_{loss}(kW)$	73.767	31.9649	30.1927	33.6443	30.0164	27.6494
	$V_{min}(p.u)$	0.9247	0.9678	0.9647	0.9673	0.9826	0.9832
	VSI(p.u)	0.7305	0.8694	0.8655	0.8748	0.924	0.9264
	$P_{load}(MW)$	3.45	0.9618	1.6865	1.4515	1.1232	1.4584
Industrial Load	Location	...	18, 60, 63	8, 13, 62	62, 51, 17	58, 64, 68	16, 61, 60
	Size (MW)	...	0.9861, 0.6573, 0.5394	0.9559, 0.7189, 1	1, 0.7399, 0.3653	0.9444, 0.8343, 0.7527	0.514, 0.7138, 1
	$P_{loss}(kW)$	171.4316	49.24	43.8302	44.6244	39.7631	30.7966
	$Q_{loss}(kW)$	79.0819	25.3316	23.1395	23.9029	21.4549	18.2978
	$V_{min}(p.u)$	0.9197	0.9692	0.9681	0.9648	0.9838	0.9864
	VSI(p.u)	0.7135	0.8782	0.8739	0.862	0.9286	0.943
	$P_{load}(MW)$	3.7679	1.5851	1.0931	1.6627	1.2365	1.5401
	Location	...	62, 67, 10	67, 57, 63	62, 57, 63	68, 63, 62	60, 16, 61
	Size (MW)	...	0.9479, 0.9535, 0.5762	0.9984, 0.7229, 0.8594	0.8763, 0.6872, 0.7251	0.5084, 0.7894, 1	1, 0.4909, 0.583

Table 4.9

Comparative Analysis of MAQiEA with Other Algorithms (Contd...)

		Base Case	GA	PSO	GSA	ALO	MAQiEA
Residential Load	Location	...	61, 12, 23	68, 57, 61	60, 63, 56	68, 62, 63	60, 61, 17
	Size (MW)	...	1, 0.52, 0.5	0.87, 1, 0.88	0.7964, 0.9312, 0.9657	0.3607, 0.9326, 0.9351	1, 0.7733, 0.5084
	$P_{loss}(kW)$	164.9382	56.1564	48.5161	54.8719	45.0331	39.6854
	$Q_{loss}(kW)$	76.3458	27.9655	25.0357	27.7732	24.1137	22.1611
	$V_{min}(p.u)$	0.9218	0.9674	0.978	0.9746	0.9803	0.9834
	VSI(p.u)	0.721	0.8729	0.9091	0.902	0.9233	0.9332
	$P_{load}(MW)$	3.6331	1.1555	1.0524	1.3445	1.3353	1.5592
	Location	...	65, 59, 22	63, 51, 23	11, 8, 62	62, 68, 63	61, 16, 60
	Size (MW)	...	1, 0.8524, 0.6827	1, 0.9492, 0.6685	0.9291, 0.7764, 0.8329	0.6022, 0.8156, 0.8462	0.5774, 0.4854, 0.9666
Commercial Load	$P_{loss}(kW)$	157.0083	58.0609	53.1034	56.9336	47.7983	43.1541
	$Q_{loss}(kW)$	72.9508	30.0867	27.2928	28.6663	25.1554	23.6074
	$V_{min}(p.u)$	0.9242	0.9961	0.9726	0.9662	0.984	0.9842
	VSI(p.u)	0.7289	0.9615	0.8892	0.8697	0.9312	0.9367
	$P_{load}(MW)$	3.5312	0.9961	0.9135	0.9662	1.2672	1.5018
	Location	...	62, 56, 61	68, 62, 18	62, 66, 61	63, 26, 61	61, 16, 60
	Size (MW)	...	0.8412, 0.6739, 0.9415	0.7077, 0.7578, 0.6394	0.8043, 0.8651, 0.7429	0.811, 0.3389, 1	0.7482, 0.5317, 1
Mixed Load	$P_{loss}(kW)$	208.8178	71.3398	66.3952	65.7329	64.6133	62.3229
	$Q_{loss}(kW)$	95.0525	35.2727	33.5744	32.821	32.7734	31.7996
	$V_{min}(p.u)$	0.9121	0.9736	0.9727	0.9758	0.9826	0.9808
	VSI(p.u)	0.6913	0.8986	0.8898	0.9054	0.9281	0.9242
	$P_{load}(MW)$	3.7415	1.2849	1.6366	1.3292	1.5916	1.4616

GSA produce active power loss of 63.5kW and 66.99kW. GA has minimum power loss reduction of 71.1685kW and 31.9649kVAr with locations 27, 64 and 49 with optimal capacity 717kW, 972kW and 800kW. Maximum percentage power loss reduction is obtained with MAQiEA as 52.6348kW and 27.6494kVAr with voltage profile improvement of 0.9832 p.u. MAQiEA has maximum improvement in VSI in comparison with all other algorithms, minimum VSI of MAQiEA is 0.9264p.u, ALO has minimum VSI of 0.924p.u followed by GSA with 0.8748p.u, followed by GA with 0.8694p.u and PSO with minimum VSI of 0.8655p.u. Improvement in voltage profile is observed, after installing DG, proposed algorithm has maximum improvement in voltage profile of 0.9832p.u and ALO has improvement in voltage profile of 0.9826p.u.

For industrial load model, the performance of MAQiEA is better in all instances i.e., real power loss, reactive power loss, voltage profile and VSI. The tabulated results demonstrate that MAQiEA has high reduction in power loss in comparison with other algorithms. Overall real and reactive power losses incurred in the system after implementing DG is 30.7966kW and 18.2978kVAr at location 16, 61, and 60 with optimal capacity 514kW, 714kW and 1MW with minimum VSI and voltage profile of 0.943p.u and 0.9864p.u respectively. ALO, GSA, PSO and GA has minimum power loss of 39.7631kW, 44.62kW, 43.83kW and 49.24kW. Minimum improvement in voltage profile is observed in GA of 0.9692p.u, whereas GSA has minimum VSI of 0.862p.u.

Similarly for residential load and commercial load, the power losses obtained after installing DG with MAQiEA is 39.6854kW and 43.1541kW. GA has power loss reduction of 56.1564kW and 58.0609kVAr. For residential load, optimal location and capacity of DG with MAQiEA is 60, 16, 61 and 1MW, 491kW and 583kW. For commercial load, optimal placement and sizing of DG with the proposed algorithm is 61, 16, 60 and 578kW, 485kW and 967kW respectively. MAQiEA has maximum reduction in power loss as compared with other algorithm for residential and commercial load models. For residential load model, ALO has power loss reduction of 45.033kW and 24.113kVAr with locations 68, 63, 62 and 508kW, 789kW and 1MW respectively. Similarly for commercial load model, ALO has active and reactive power loss reduction of 47.7983kW and 25.1554kVAr with optimal placement and sizing of DG 62, 68, 63 and 602kW, 816kW and 846kW with minimum voltage profile of 0.984kW and VSI of 0.9312p.u. For mixed load model, integration of all voltage dependent load models is considered. The load at every bus is given in appendix. MAQiEA has maximum improvement in voltage profile and VSI of 0.9808p.u and 0.9242p.u. Whereas other algorithms such as ALO, GSA, PSO and GA has minimum voltage profile of 0.9823p.u, 0.9758p.u, 0.9727p.u, and 0.9736p.u. Minimum VSI of ALO, GSA, PSO and GA are 0.9281p.u, 0.9054p.u, 0.8898p.u, and 0.8986p.u respectively. Minimum power losses are obtained with MAQiEA of 62.3229kW and 31.7996kVAr with location 61, 16, 60 and capacities of 748kW, 532kW and 1MW respectively.

Optimal location and capacity of DG not only reduces the power losses but also improves the voltage profile. Improvement in voltage profile for constant power

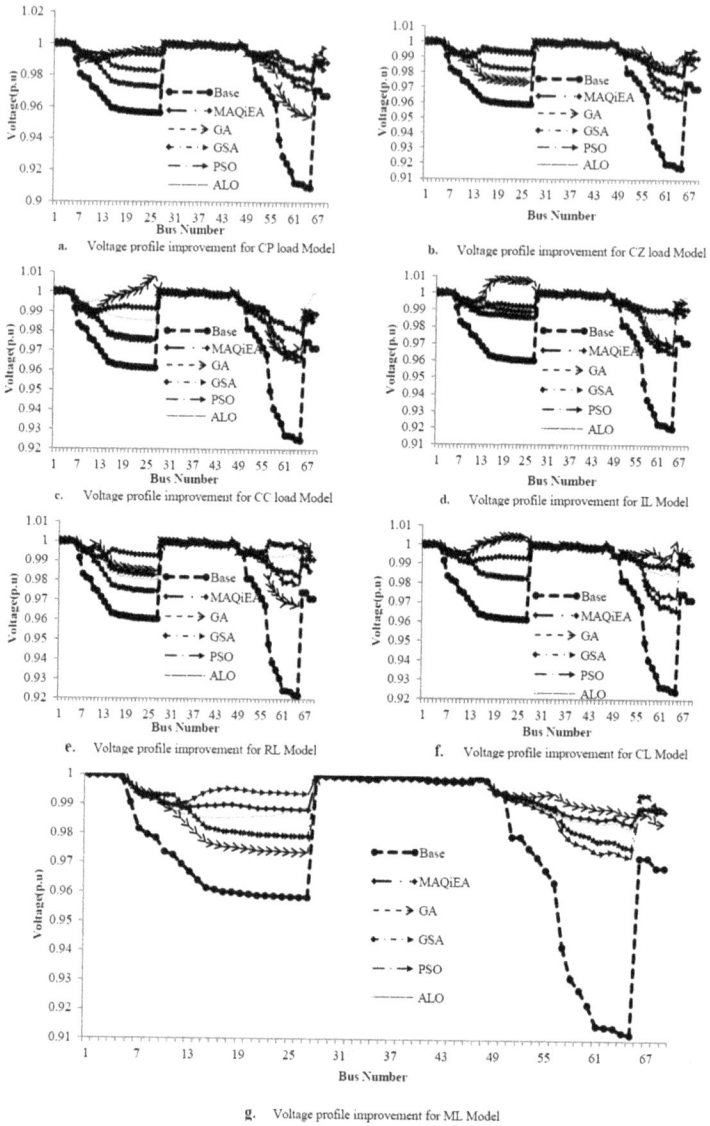

a.   Voltage profile improvement for CP load Model

b.   Voltage profile improvement for CZ load Model

c.   Voltage profile improvement for CC load Model

d.   Voltage profile improvement for IL Model

e.   Voltage profile improvement for RL Model

f.   Voltage profile improvement for CL Model

g.   Voltage profile improvement for ML Model

Figure 4.11: Voltage profile improvement for test bus system.

load, constant impedance load, constant current load, industrial load, residential load, commercial load and mix load models are shown in Figure 4.11 with all algorithms including base case. Similarly for improvement in VSI for constant power load, constant impedance load, constant current load, industrial load, residential load,

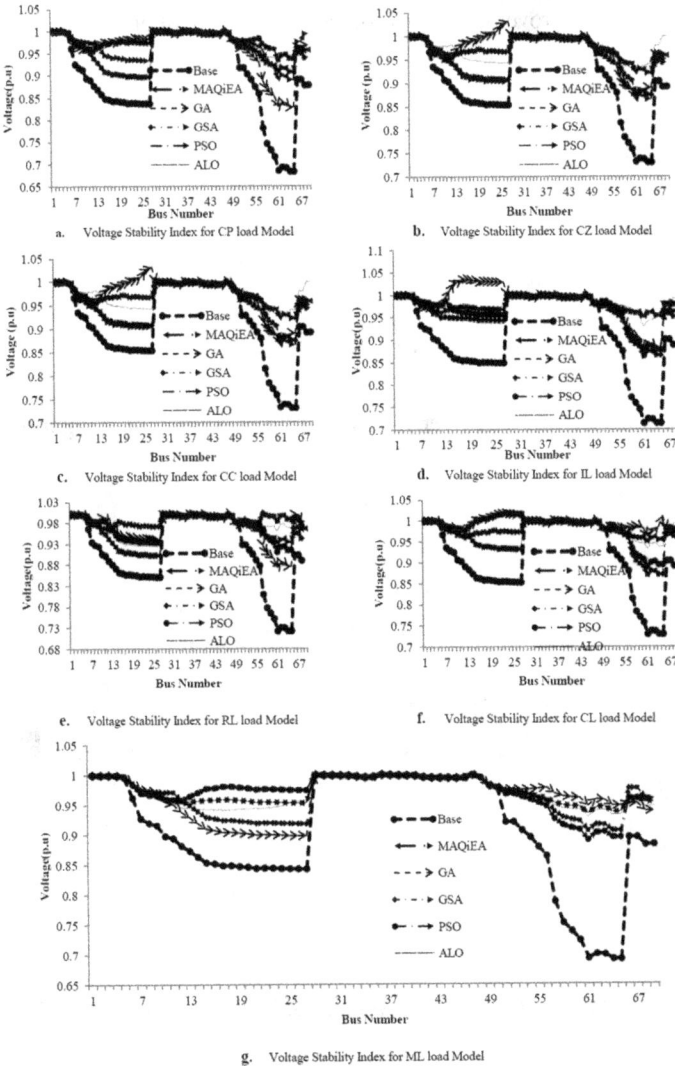

a. Voltage Stability Index for CP load Model

b. Voltage Stability Index for CZ load Model

c. Voltage Stability Index for CC load Model

d. Voltage Stability Index for IL load Model

e. Voltage Stability Index for RL load Model

f. Voltage Stability Index for CL load Model

g. Voltage Stability Index for ML load Model

Figure 4.12: Voltage Stability Index for test bus system

commercial load and mix load models are shown in Figure 4.12 with all algorithms including base case.

Figure 4.13 shows the overall power loss comparison of MAQiEA with ALO, GSA, PSO, GA and base case. Proposed algorithm has high reduction in power loss for all load models including constant power load, Figure 4.14 shows the overall improvement in voltage profile of MAQiEA with all different load models. Figure 4.15 shows the overall improvement in VSI of MAQiEA with all different load models.

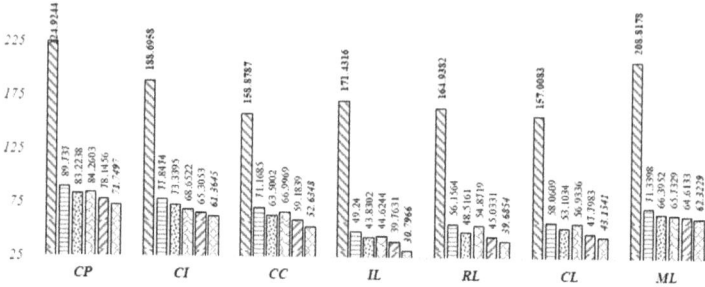

Figure 4.13: Comparison of power loss with MAQiEA with other algorithms for all load models.

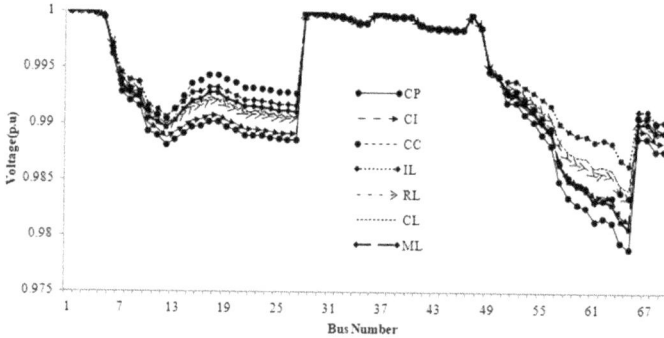

Figure 4.14: Voltage Profile improvement of MAQiEA for all Load Models.

Discussions:

Distribution network has major power loss percentage in power system network as compared with generation and transmission system. Distributed generators are normally employed in DN to reduce the power losses. Majority of researchers have implemented DG in DN with a CP load model. However, it is well known that load at DN varies from time to time. In this study, an investigation has been performed with DG on different load models to reduce the losses. Voltage-dependent load are used in this study. CP load model is generally independent of voltage, which doesn't vary with voltage. Most of the loads used by consumers at load centers are dependents on

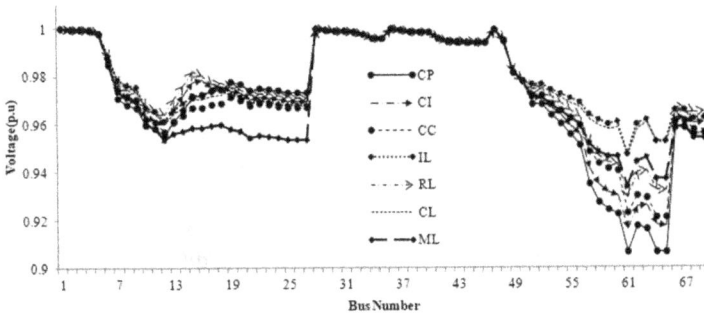

Figure 4.15: Voltage Stability Index of MAQiEA for all Load Models.

voltage. So it is necessary to know the effect of DG on VDLM. CC load and CZ load models varies linearly and square of the voltage. In addition, some practical loads which are majorly used in distribution network are also considered viz., RL, IL and CL models. A class of ML is considered which combines all the voltage dependent loads. Seven different cases are used to study the effect of DG. If a CP load model is used, the total load on the test bus system is independent of voltage. Total load used in the distribution network didn't dependent on voltage. If a CZ load model is used, the total load on the test bus system linearly varies with voltage. Total load used in the test bus linearly varies with voltage. Similarly, in case of CC load model, total load on the test bus system varies with square of the voltage. In case of practical loads, if an IL is used, total load used by the test bus system is only industrial load. Similarly, for CL and RL the total load on the test bus system is commercial load and residential load. Combination of all load models including CP load model is used in ML model. Tables 4.10 and 4.11 shows the data of benchmark test bus system.

Multiple DGs are used in this study to reduce the power losses. Multiple DGs have high reduction in power loss as compared with single DG implementation. In case of single DG, one DG with high power rating is used to reduce the losses, whereas multiple DG uses small power rating in comparison with single DG rating. Optimal placement and sizing of DG is complex combinatorial optimization problem. Numerical, analytical and meta-heuristic algorithms are used to reduce the losses. For large test bus system, metaheuristics has high reduction in power losses and low computation time in comparison with analytical and numerical methods. Many metaheuristic techniques are implemented for this important optimization problem. Normally, evolutionary algorithms suffer from slow convergence, stagnation of choice of parameter and premature convergence. An evolutionary algorithm which overcomes the limitations in EA is QiEA. In this study, a new and efficient quantum-inspired evolutionary algorithm is used to find the optimal location and size of DG, which is termed as Multi-partite Adaptive Quantum inspired Evolutionary Algorithm. MAQiEA uses probabilistic approach with Q-bits. MAQiEA is an updated version of AQiEA, which was an improvement over QiEA. MAQiEA didnt require any additional operator to avoid premature convergence. QiEA uses single Q-bit whereas AQiEA used two Q-bits. Q-gates are used in QiEA to move the system towards convergence whereas MAQiEA used a Multi-partite Adaptive Crossover operator for better convergence. AQiEA has used three rotation strategies to move the search towards better solutions, which were bipartite Rotation towards Best, Rotation towards Better and Rotation away from Worse. Whereas MAQiEA uses improved probabilistic rotation around better & multi-partite rotation away from worse rotation strategies which provides for relatively better exploration and exploitation, in addition to Rotation towards Best Strategy in AQiEA. MAQiEA has high robustness and has better exploitation and exploration of search space in comparison with AQiEA as shown by test results. Wilcoxon signed rank test has been used to arrive at best design of MAQiEA amongst various alternatives. It has been observed from tabulated results that for voltage dependent load models that location of DG is fixed for all load models except constant power and constant current load. The load on the system is varying exponentially with voltage level at the node in case of IL, CL and RL. The optimal location of DG for practical loads is fixed for MAQiEA but other algorithms have different optimal location for different load models. The robustness of the proposed algorithm is very high. The results of simulated experiments in the tables demonstrate that MAQiEA is performing better in comparison with other algorithms (GA, GSA, PSO and ALO).

4.7   CONCLUSIONS

Minimization of power loss in a DN is one of the challenging areas of research for the distribution utilities. In recent times, power losses are reduced by implementing DGs into distribution network. However, majority of research has been done on this important optimization problem with CP load model. Majority of consumers at load center uses VDLMs such as CZ, CC, IL, RL and CL, whereas CP load model is independent of voltage. If the optimal placement and capacity of DG with CP load model is used on practical distribution system, it induces high power losses and poor voltage regulation in the system. In this study, an investigation has been performed to reduce the losses in the distribution system with DG for different VDLMs. Optimal location and capacity of DG is a difficult non differentiable, non linear, complex combinatorial optimization problem. A Multipartite Adaptive Quantum-inspired Evolutionary Algorithm is proposed for optimal location and sizing of DG. MAQiEA uses probabilistic approach with Q-bits, and is an updated version of AQiEA that has introduced two Q-bits per solution vector and entanglement inspired adaptive crossover operator. MAQiEA has introduced a Multipartite Adaptive Crossover operator as a variation operator for better convergence. The effectiveness of MAQiEA is tested on standard IEEE benchmark test bus system. Tabulated result shows the effectiveness of the proposed algorithm as compared with other algorithms.

REFERENCES

1. Sorensen, K., Sevaux, M., and Glover, F. (2017). A history of metaheuristics. arXiv preprint arXiv:1704.00853.
2. Blum, C., and Roli, A. (2008). Hybrid metaheuristics: An introduction. In Hybrid Metaheuristics (pp. 1–30). Springer, Berlin, Heidelberg.
3. Singh, B., and Sharma, J. (2017). A review on distributed generation planning. Renewable and Sustainable Energy Reviews, 76, 529–544.
4. Han, K. H., and Kim, J. H. (2002). Quantum-inspired evolutionary algorithm for a class of combinatorial optimization. IEEE Transactions on Evolutionary Computation, 6(6), 580–593.
5. Zhang, G. (2011). Quantum-inspired evolutionary algorithms: A survey and empirical study. Journal of Heuristics, 17(3), 303–351.
6. Mani, A., and Patvardhan, C. (2009, May). A novel hybrid constraint handling technique for evolutionary optimization. In 2009 IEEE Congress on Evolutionary Computation (pp. 2577–2583). IEEE.
7. Manikanta, G., Mani, A., Singh, H. P., and Chaturvedi, D. K. (2016, September). Placing distributed generators in distribution system using adaptive quantum inspired evolutionary algorithm. In 2016 Second International Conference on Research in Computational Intelligence and Communication Networks (ICRCICN) (pp. 157–162). IEEE.
8. Manikanta, G., Mani, A., Singh, H. P., and Chaturvedi, D. K. (2019). Adaptive quantum-inspired evolutionary algorithm for optimizing power losses by dynamic load allocation on distributed generators. SJEE, 16(3), 325–357.
9. Manikanta, G., Mani, A., Singh, H. P., and Chaturvedi, D. K. (2019). Distribution Network Reconfiguration using Adaptive quantum-inspired evolutionary algorithm. International Conference on Recent innovation in Electrical Electronics and Communication Engineering (ICRIEECE-2018) at School of Electrical Engineering, Kalinga Institute of Industrial Technology (KIIT), Bhubaneswar, India
10. Manikanta, G., Mani, A., Singh, H. P., and Chaturvedi, D. K. (2018, December). Minimization of Power Losses in Distribution System with Variation in Loads Using Adaptive Quantum inspired Evolutionary Algorithm. In 2018 4th International Conference on Computing Communication and Automation (ICCCA) (pp. 1–6). IEEE.
11. Manikanta, G., Mani, A., Singh, H. P., and Chaturvedi, D. K. (2018, October). Distribution Network Reconfiguration with Different Load Models using Adaptive Quantum inspired Evolutionary Algorithm. In 2018 International Conference on Sustainable Energy, Electronics, and Computing Systems (SEEMS) (pp. 1–7). IEEE.
12. Mani, A., and Patvardhan, C. (2012). An improved model of ceramic grinding process and its optimization by adaptive Quantum inspired evolutionary algorithm. International Journal of Simulations: Systems Science and Technology, 11(6), 76–85.
13. Manikanta, G., Mani, A., Singh, H. P., and Chaturvedi, D. K. (2017). DG and Capacitor Placement in Distribution system considering Cost and Benefits using AQiEA, National System Conference, DEI, Agra.
14. Manikanta, G., Mani, A., Singh, H. P., and Chaturvedi, D. K. (2016, November). Sitting and sizing of capacitors in distribution system using adaptive quantum inspired evolutionary algorithm. In 2016 7th India International Conference on Power Electronics (IICPE) (pp. 1–6). IEEE.
15. Manikanta, G., Mani, A., Singh, H. P., and Chaturvedi, D. K. (2019). Simultaneous placement and sizing of DG and capacitor to minimize the power losses in radial

distribution network. In Soft computing: Theories and applications (pp. 605–618). Springer, Singapore.

16. Sultana, U., Khairuddin, A., Mokhtar, A. S., Qazi, S. H., and Sultana, B. (2017). An optimization approach for minimizing energy losses of distribution systems based on distributed generation placement. Jurnal Teknologi, 79(4), 87–96.

17. Manikanta, G., Mani, A., Singh, H. P., and Chaturvedi, D. K. (2017, November). Minimization of power losses in distribution system using symbiotic organism search algorithm. In 2017 IEEE PES Asia-Pacific Power and Energy Engineering Conference (APPEEC) (pp. 1–6). IEEE.

18. Seyedali M., Gandomi A., Mirjalili S., Saremi S., Faris H., and Mirjalili S. (2017) Salp Swarm Algorithm: A bio-inspired optimizer for engineering design problems. Advances in Engineering Software, 114:163–191.

19. Lopes, J. P., Hatziargyriou, N., Mutale, J., Djapic, P., and Jenkins, N. (2007). Integrating distributed generation into electric power systems: A review of drivers, challenges and opportunities. Electric Power Systems Research, 77(9), 1189–1203.

20. Ackermann, T., Andersson, G., and Sder, L. (2001). Distributed generation: A definition. Electric Power Systems Research, 57(3), 195–204.

21. Viral, R., and Khatod, D. K. (2012). Optimal planning of distributed generation systems in distribution system: A review. Renewable and Sustainable Energy Reviews, 16(7), 5146–5165.

22. Babu, P. V., Singh, S., and Singh, S. P. (2017, July). Distributed generators allocation in distribution system. In 2017 IEEE Power & Energy Society General Meeting (pp. 1–5). IEEE.

23. Prakash, P., and Khatod, D. K. (2016, July). An analytical approach for optimal sizing and placement of distributed generation in radial distribution systems. In 2016 IEEE 1st International Conference on Power Electronics, Intelligent Control and Energy Systems (ICPEICES) (pp. 1–5). IEEE.

24. Vatani, M., Alkaran, D. S., Sanjari, M. J., and Gharehpetian, G. B. (2016). Multiple distributed generation units allocation in distribution network for loss reduction based on a combination of analytical and genetic algorithm methods. IET Generation, Transmission and Distribution, 10(1), 66–72.

25. Al Ameri, A., Nichita, C., Abbood, H., and Al Atabi, A. (2015, March). Fast Estimation Method for Selection of Optimal Distributed Generation Size Using Kalman Filter and Graph Theory. In 2015 17th UKSim-AMSS International Conference on Modelling and Simulation (UKSim) (pp. 420–425). IEEE.

26. Mahmoud, Karar, Yorino, Naoto and Ahmed, Abdella. (2015). Power loss minimization in distribution systems using multiple distributed generations. IEEJ Transactions on Electrical and Electronic Engineering, 10: 521–526, 2015. doi:10.1002/tee.22115.

27. Khoa, T. H., Nallagownden, P., Baharudin, Z., and Dieu, V. N. (2017, November). One rank cuckoo search algorithm for optimal placement of multiple distributed generators in distribution networks. In TENCON 2017-2017 IEEE Region 10 Conference (pp. 1715–1720). IEEE.

28. Guo, Y., Gao, H., Wang, J., Wu, Z., and Han, C. (2015, November). Analysis of distributed generation effect on system losses in distribution network. In 2015 5th International Conference on Electric Utility Deregulation and Restructuring and Power Technologies (DRPT) (pp. 1998–2002). IEEE.

29. Mistry, K. (2016, March). MSFL based determination of optimal size and location of distributed generation in radial distribution system. In 2016 International Conference on

Electrical, Electronics, and Optimization Techniques (ICEEOT) (pp. 530–535). IEEE.

30. de Carvalho, T. L. A., and Ferreira, N. R. (2018, May). Optimal allocation of distributed generation using ant colony optimization in electrical distribution system. In 2018 Simposio Brasileiro de Sistemas Eletricos (SBSE) (pp. 1–6). IEEE.

31. Vizhiy, S. A., and Santhi, R. K. (2016, March). Biogeography based optimal placement of distributed generation units in distribution networks: Optimal placement of distributed generation units. In 2016 International Conference on Electrical, Electronics, and Optimization Techniques (ICEEOT) (pp. 2245–2250). IEEE.

32. Ali, E. S., Abd Elazim, S. M., and Abdelaziz, A. Y. (2017). Ant Lion Optimization Algorithm for optimal location and sizing of renewable distributed generations. Renewable Energy, 101, 1311–1324, 2017.

33. Mahajan, S., and S. Vadhera, S. (2016, March). Optimal sizing and deploying of distributed generation unit using a modified multiobjective Particle Swarm Optimization. In 2016 IEEE 6th International Conference on Power Systems (ICPS) (pp. 1–6). IEEE.

34. Saha, S., and Mukherjee, V. (2016). Optimal placement and sizing of DGs in RDS using chaos embedded SOS algorithm. IET Generation, Transmission and Distribution, 10(14), 3671–3680.

35. Behera, S. R., and Panigrahi, B. K. (2019). A multi objective approach for placement of multiple DGs in the radial distribution system. International Journal of Machine Learning and Cybernetics, 10(8), 2027–2041.

36. Saini, S., and Kaur, G. (2016, December). Real power loss reduction in distribution network through Distributed Generation integration by implementing SPSO. In 2016 International Conference on Electrical Power and Energy Systems (ICEPES) (pp. 35–40). IEEE.

37. Bansal, A., and Singh, S. (2016). Optimal allocation and sizing of distributed generation for power loss reduction.

38. Mer, D. K., and Patel, R. R. (2016, March). The concept of distributed generation and the effects of its placement in distribution network. In 2016 International Conference on Electrical, Electronics, and Optimization Techniques (ICEEOT) (pp. 3965–3969). IEEE.

39. Jamian, J. J., Mustafa, M. W., Mokhlis, H., Baharudin, M. A., and Abdilahi, A. M. (2014). Gravitational search algorithm for optimal distributed generation operation in autonomous network. Arabian Journal for Science and Engineering, 39(10), 7183–7188.

40. Zhang, J., and Bo, Z. Q. (2010, September). Research of the impact of distribution generation on distribution network loss. In 45th International Universities Power Engineering Conference UPEC2010 (pp. 1–4). IEEE, 2010.

41. Hasibuan, A., Masri, S., and Othman, W. A. F. W. B. (2018, February). Effect of distributed generation installation on power loss using genetic algorithm method. In IOP Conference Series: Materials Science and Engineering (Vol. 308, No. 1, p. 012034). IOP Publishing.

42. Ang, S., Leeton, U., Chayakulkeeree, K., and Kulworawanichpong, T. (2018). Sine cosine algorithm for optimal placement and sizing of distributed generation in radial distribution network. GMSARN International Journal, 12, 202–212.

43. Roy, N. K., Hossain, M. J., and Pota, H. R. (2011, September). Effects of load modeling in power distribution system with distributed wind generation. In AUPEC 2011 (pp. 1–6). IEEE.

44. Angarita, O. F. B., Leborgne, R. C., Gazzana, D. D. S., and Bortolosso, C. (2015, October). Power loss and voltage variation in distribution systems with optimal allocation

of distributed generation. In 2015 IEEE PES Innovative Smart Grid Technologies Latin America (ISGT LATAM) (pp. 214–218). IEEE.

45. Divya, K., and Srinivasan, S. (2016, January). Optimal siting and sizing of DG in radial distribution system and identifying fault location in distribution system integrated with distributed generation. In 2016 3rd International Conference on Advanced Computing and Communication Systems (ICACCS) (Vol. 1, pp. 1–7). IEEE.

46. Bohre A. K., Agnihotri G. (2016). Optimal sizing and sitting of DG with load models using soft computing techniques in practical distribution system. IET Generation, Transmission & Distribution, 10(11), 2606–2621.

47. Das, S., Das, D., and Patra, A. (2016, July). Distribution network reconfiguration using distributed generation unit considering variations of load. In 2016 IEEE 1st International Conference on Power Electronics, Intelligent Control and Energy Systems (ICPEICES) (pp. 1–5). IEEE.

48. Price, W. W., Casper, S. G., Nwankpa, C. O., Bradish, R. W., Chiang, H. D., Concordia, C., ... and Wu, G. (1995). Bibliography on load models for power flow and dynamic performance simulation. IEEE Power Engineering Review, 15(2), 70.

49. Price, W. W., Taylor, C. W., and Rogers, G. J. (1995). Standard load models for power flow and dynamic performance simulation. IEEE Transactions on Power Systems, 10(CONF-940702-), 1302–1313.

50. Concordia, C., and Ihara, S. (1982). Load representation in power system stability studies. IEEE Transactions on Power Apparatus and Systems, (4), 969–977.

51. Price, W. W., Chiang, H. D., Clark, H. K., Concordia, C., Lee, D. C., Hsu, J. C., ... and Vaahedi, E. (1993). Load representation for dynamic performance analysis. IEEE Transactions on Power Systems (Institute of Electrical and Electronics Engineers);(United States), 8(2).

52. Teng, J. H. (2003). A direct approach for distribution system load flow solutions. IEEE Transactions on Power Delivery, 18(3), 882–887.

53. Garca, S., Molina, D., Lozano, M., and Herrera, F. (2009). A study on the use of non-parametric tests for analyzing the evolutionary algorithms behaviour: A case study on the CEC2005 special session on real parameter optimization. Journal of Heuristics, 15(6), 617–644.

54. Derrac, J., Garca, S., Molina, D., and Herrera, F. (2011). A practical tutorial on the use of nonparametric statistical tests as a methodology for comparing evolutionary and swarm intelligence algorithms. Swarm and Evolutionary Computation, 1(1), 3–18.

4.8   PARAMETERS OF IEEE BENCHMARK TEST BUS SYSTEM

The detailed parameters of IEEE benchmark test bus system are listed in the Tables 4.10 and 4.11.

Table 4.10
Line Parameters for IEEE 69 Bus Radial Distribution System

Branch No	From Bus	To Bus	R(ohms)	X(ohms)
1	1	2	0.0005	0.00112
2	2	3	0.0005	0.00112
3	3	4	0.0015	0.0036
4	4	5	0.0251	0.0294
5	5	6	0.366	0.1864
6	6	7	0.381	0.1941
7	7	8	0.0922	0.047
8	8	9	0.0493	0.0251
9	9	10	0.819	0.2707
10	10	11	0.1872	0.0619
11	11	12	0.7114	0.2351
12	12	13	1.03	0.34
13	13	14	1.044	0.345
14	14	15	1.058	0.3496
15	15	16	0.1966	0.065
16	16	17	0.3744	0.1238
17	17	18	0.0047	0.0016
18	18	19	0.3276	0.1083
19	19	20	0.2106	0.069
20	20	21	0.3416	0.1129
21	21	22	0.014	0.0046
22	22	23	0.1591	0.0526
23	23	24	0.3463	0.1145
24	24	25	0.7488	0.2475
25	25	26	0.3089	0.1021
26	26	27	0.1732	0.0572
27	3	28	0.0044	0.0108
28	28	29	0.064	0.1565
29	29	30	0.3978	0.1315
30	30	31	0.0702	0.0232
31	31	32	0.351	0.116
32	32	33	0.839	0.2816
33	33	34	1.708	0.5646
34	34	35	1.474	0.4873
35	3	36	0.0044	0.0108
36	36	37	0.064	0.1565
37	37	38	0.1053	0.123
38	38	39	0.0304	0.0355
39	39	40	0.0018	0.0021
40	40	41	0.7283	0.8509
41	41	42	0.31	0.3623
42	42	43	0.041	0.0478
43	43	44	0.0092	0.0116
44	44	45	0.1089	0.1373
45	45	46	0.0009	0.0012
46	4	47	0.0034	0.0084
47	47	48	0.0851	0.2083
48	48	49	0.2898	0.7091
49	49	50	0.0822	0.2011
50	8	51	0.0928	0.0473
51	51	52	0.3319	0.1114
52	9	53	0.174	0.0886
53	53	54	0.203	0.1034
54	54	55	0.2842	0.1447
55	55	56	0.2813	0.1433
56	56	57	1.59	0.5337
57	57	58	0.7837	0.263
58	58	59	0.3042	0.1006
59	59	60	0.3861	0.1172
60	60	61	0.5075	0.2585
61	61	62	0.0974	0.0496
62	62	63	0.145	0.0738
63	63	64	0.7105	0.3619
64	64	65	1.041	0.5302
65	11	66	0.2012	0.0611
66	66	67	0.0047	0.0014
67	12	68	0.7394	0.2444
68	68	69	0.0047	0.0016

Table 4.11
Load Parameters for IEEE 69 Bus Radial Distribution System

Bus No	Active Load (kW)	Reactive Load (kVAr)	Load Model
1	0	0	Substation
2	0	0	Constant Impedance Load
3	0	0	Constant Current Load
4	0	0	Constant Current Load
5	2.6	2.2	Constant power Load
6	40.4	30	Residential Load
7	75	54	Industrial Load
8	30	22	Constant power Load
9	28	19	Industrial Load
10	145	104	Constant Current Load
11	145	104	Commercial Load
12	8	5	Commercial Load
13	8	5.5	Commercial Load
14	0	0	Residential Load
15	45.5	30	Constant Current Load
16	60	35	Constant Impedance Load
17	60	35	Constant Impedance Load
18	0	0	Commercial Load
19	1	0.6	Constant power Load
20	114	81	Commercial Load
21	5	3.5	Constant Impedance Load
22	0	0	Constant Current Load
23	28	20	Industrial Load
24	0	0	Constant Impedance Load
25	14	10	Constant power Load
26	14	10	Commercial Load
27	26	18.6	Constant power Load
28	26	18.6	Residential Load
29	0	0	Residential Load
30	0	0	Commercial Load
31	0	0	Residential Load
32	14	10	Commercial Load
33	19.5	14	Commercial Load
34	6	4	Constant power Load
35	26	18.55	Constant Impedance Load
36	26	18.55	Constant Current Load
37	0	0	Commercial Load
38	24	17	Constant power Load
39	24	17	Constant Current Load
40	1.2	1	Commercial Load
41	0	0	Constant power Load
42	6	4.3	Constant Current Load
43	0	0	Commercial Load
44	39.22	26.3	Industrial Load
45	39.22	26.3	Commercial Load
46	0	0	Industrial Load
47	79	56.4	Constant power Load
48	384.7	274.5	Residential Load
49	384.7	274.5	Constant power Load
50	40.5	28.3	Residential Load
51	3.6	2.7	Constant Impedance Load
52	4.35	3.5	Constant Impedance Load
53	26.4	19	Constant Impedance Load
54	24	17.2	Industrial Load
55	0	0	Commercial Load
56	0	0	Industrial Load
57	0	0	Residential Load
58	100	72	Constant Current Load
59	0	0	Constant Current Load
60	1244	888	Constant power Load
61	32	23	Constant power Load
62	0	0	Constant Impedance Load
63	227	162	Constant Current Load
64	59	42	Industrial Load
65	18	13	Constant power Load
66	18	13	Industrial Load
67	28	20	Constant Impedance Load
68	28	20	Constant Impedance Load

5 Quantum-Inspired Manta Ray Foraging Optimization Algorithm for Automatic Clustering of Color Images

5.1 INTRODUCTION

Clustering or cluster analysis can be defined as a process of discovering the underlying structure of a data set by partitioning the entire data into two or more groups. In this process, similar data points are kept together in the same group and dissimilar data points are kept separate.

The extensive use of *clustering* can be applied in the field of data mining, engineering, economics, sociology, biology, and physics [1][2]. In order to deal with the clustering problems, several approaches have been proposed so far, which includes hierarchical clustering, non-hierarchical clustering, fuzzy clustering, artificial neural network-based clustering and evolutionary approach-based clustering [1, 2] to name a few. In the literature, a variety of clustering methods have been developed so far since the past few years. The prerequisite for most of the existing methods is that they must have the knowledge of apposite number of clusters beforehand. In most occasions, the dataset may suffer from insufficient and inappropriate knowledge about the data, that makes the functioning of clustering algorithm, a challenging and tedious task. To cope up with this limitation, few *automatic clustering* techniques have been already developed by several researchers [8][9][10][11].

In the recent years, *metaheuristic* algorithms have been considered a good choice for solving several kinds of optimization problems. They are able to provide an appropriate solution for different simple and complex optimization problems within a short time frame. Some popular metaheuristic algorithms may include Genetic Algorithm [7], Particle Swarm Optimization [8], Differential Evaluation [9], etc. Though, the nature-inspired metaheuristic algorithms are capable to solve a problem very quickly, still they may suffer from premature convergence. In order to handle this situation, several new approaches can be adopted efficiently and effectively. These approaches may include introducing new parameters in the existing algorithm, hybridizing more than one algorithm or even incorporating the features of quantum computing into an existing algorithm. In this regard, quantum-inspired metaheuristic algorithms have achieved a remarkable efficiency with reference to

DOI: 10.1201/9781003283294-5

fast computation without getting stuck in the local optima [10][11][28][13][14][15] [34][17].

In this paper, an effort has been made to introduce a new *quantum-inspired* algorithm by coupling the features of quantum computing and a newly developed *metaheuristic* algorithm, called Manta Ray Foraging Optimization [18] algorithm.

The main features of the proposed work are stated below.

A new computational methodology has been developed for *automatic clustering* of color image by incorporating the features of quantum computing into a *metaheuristic*, called Manta Ray Foraging Optimization [18] algorithm. This algorithm is able to identify the optimal number of cluster effectively and efficiently without getting stuck in local optima.

The exploration and exploitation strategies in the search space have been achieved by using quantum *rotation* gate and *Pauli-X* gate.

The implementation process of the proposed algorithm is simple.

The parameters of the proposed algorithm have been tunes by the Sobol's *sensitivity analysis* [19][20][21] to enhance its efficacy.

The rest of the paper has been organized as follows: A brief review has been presented in Section 5.2. Section 5.3 presents a brief idea about quantum computing. Section 5.4 describes the validity measurement process of the *clustering* algorithms. A brief overview of the Manta Ray Foraging Optimization algorithm has been presented in Section 5.5. Section 5.6 elaborates the working principle of the proposed method. The experimental details and analysis of results have been presented in Section 5.7. Finally, Section 5.8 presents a brief conclusion with possible future scope.

5.2   LITERATURE REVIEW

Clustering algorithms are basically categorized into two types, viz., partitional *clustering* and hierarchical clustering [22][1][2]. In case of partitional clustering, the entire dataset is decomposed into a set of disjoint clusters. Few examples of partitional clustering methods are the K-means algorithm, its variations [23][24], and the expectation maximization (EM) algorithm [25] to name a few. In contrast, a tree structure is generated during the hierarchical clustering, in which, each cluster represents the partition of the data set. Few representative examples of hierarchical clustering are single-linkage, complete-linkage and average-linkage methods [1][26][27] to name a few. An extensive survey of various clustering methods is presented in [2].

Nowadays, any kind of simple or complex optimization problems can be easily solved with the help of nature-inspired metaheuristics algorithms. The nature-inspired metaheuristics algorithms basically accumulate the social behavior of some creatures, viz., birds, fish, dolphins, ants, lions, wolves, honey bees, fireflies, crow, bat, etc. In order to solve any kinds of complex problems, they use their intelligence to learn from their surroundings and act accordingly. Some researchers have been motivated to design some nature-inspired meta-heuristic algorithms to solve various kinds of complex optimization problems by assimilating the intelligent behavior of the social creatures. Examples of some efficient meta-heuristic algorithms may include Genetic algorithm [7], Differential Evaluation [9], Particle Swarm Optimization [8], Ant Colony Optimization [28], Bat Optimization [29], Bacterial Foraging algorithm [30], Firefly algorithm [31], Cuckoo Search [32], and Crow Search algorithm [33][34] to name a few. By identifying the problem of automatic clustering, several nature-inspired metaheuristics algorithms have been developed so far, which are available in the literature [8][9][10][11]. The nature-inspired *metaheuristic* algorithms are capable to solve any kind of complex optimization problems within a short time frame; in addition, they are also capable to provide an optimal or near optimal solution. In spite of these capabilities, they may suffer from premature convergence. In order to overcome this situation, the quantum-inspired framework has been combined with several meta-heuristic approaches. These quantum-inspired meta-heuristic algorithms can be efficiently and effectively applied to solve various types of optimization problems, viz., task scheduling on distributed systems [23], combinational optimization problems [24][17], multi-cast routing problems in wireless mesh networks [37], multi-level thresholding problems [38][39], image analysis [15], mathematical function optimization [40][41], and *automatic clustering* [10][11][28][13][14][34] to name a few.

5.3   FUNDAMENTALS OF QUANTUM COMPUTING

In quantum computing, the smallest processing element can be referred as quantum bit or qubit, which is analogous to binary bit in classical computer. In a classical computer, a single register can store only one state of information, such as zero (0) or one(1) at any point of time, but for quantum computer, a single register is capable to store the states like $|0\rangle$ and $|1\rangle$ and their superposition state $|\Psi\rangle$ at a time. The superposition state, $|\Psi\rangle$ is basically a combination of the states $|0\rangle$ and $|1\rangle$, which can be represented as follows:

$$|\Psi\rangle = \alpha|0\rangle + \beta|1\rangle \tag{5.1}$$

where α and β represent the probability amplitudes of the corresponding states, respectively. The following normalization condition should be satisfied by them.

$$|\alpha|^2 + |\beta|^2 = 1 \tag{5.2}$$

Here, for the purpose of quantum measurement, the superposition state $|\Psi\rangle$ is collapsed either to $|0\rangle$ or to $|1\rangle$ by satisfying the following equation.

$$|\Psi\rangle = \begin{cases} |0\rangle & \text{if } |\alpha|^2 > |\beta|^2, \\ |1\rangle & \text{Otherwise} \end{cases} \tag{5.3}$$

The processing capabilities of quantum-based algorithms depend upon the quantum logic gates, which are basically congregation of several hardware devices. Quantum gates are reversible in nature. These gates can be used in different ways for developing quantum-based algorithms. Some popular *quantum gates* are Hadamard gate, Phase Shift gate, Controlled gate, Toffoli gate, Fredkin gate, Rotation gate, Pauli gate, etc [39]. In this paper, two very useful gates, viz., Rotation gate and Pauli-X gate [28][13] have been incorporated with the classical MRFO to develop proposed QIMRFO.

5.3.1 ROTATION GATE

The superposition state, $|\Psi\rangle$ as presented in Equation (5.1), can also be written as follows:

$$|\Psi\rangle = \cos\theta\,|0\rangle + \sin\theta\,|1\rangle \tag{5.4}$$

where the values of $\cos^2\theta$ and $\sin^2\theta$ represent the probability amplitudes of the states $|0\rangle$ and $|1\rangle$, respectively, and in this case, $\begin{bmatrix}\cos\theta\\\sin\theta\end{bmatrix}$ represents a single qubit. Mathematically, the quantum *rotation* gate can be represented by the following matrix.

$$U(\delta) = \begin{bmatrix}\cos\delta & -\sin\delta\\\sin\delta & \cos\delta\end{bmatrix} \tag{5.5}$$

where δ is the small rotational angle that can be applied to $\begin{bmatrix}\cos\theta\\\sin\theta\end{bmatrix}$ to produce new

qubit, $\begin{bmatrix}\cos'\theta\\\sin'\theta\end{bmatrix}$. The process to update quantum states can be accomplished by using the quantum *rotation* gate as given by

$$\begin{bmatrix}\cos'\theta\\\sin'\theta\end{bmatrix} = U(\delta)\begin{bmatrix}\cos\theta\\\sin\theta\end{bmatrix} = \begin{bmatrix}\cos\delta & -\sin\delta\\\sin\delta & \cos\delta\end{bmatrix}\begin{bmatrix}\cos\theta\\\sin\theta\end{bmatrix} = \begin{bmatrix}\cos(\theta+\delta)\\\sin(\theta+\delta)\end{bmatrix} \tag{5.6}$$

5.3.2 PAULI-X GATE

The Pauli-X gate is a quantum equivalent to NOT gate, which acts on a single *qubit* to invert its value. This can be defined as follows:

$$X = \begin{bmatrix}0 & 1\\1 & 0\end{bmatrix} \tag{5.7}$$

Mathematically, one qubit state is converted to other by using *Pauli-X* gate as follows:

$$X\,|\Psi\rangle = X\begin{bmatrix}\alpha\\\beta\end{bmatrix} = \begin{bmatrix}0 & 1\\1 & 0\end{bmatrix}\begin{bmatrix}\alpha\\\beta\end{bmatrix} = \begin{bmatrix}\beta\\\alpha\end{bmatrix} \tag{5.8}$$

The computational details of the Quantum Rotation gate and the *Pauli-X* gate have been elaborately described in [28][13].

5.4 VALIDITY MEASUREMENT OF CLUSTERING

In this paper, Pakhira Bandyopadhyay Maulik (PBM) index [42] has been used as a cluster validity index to identify the appropriate number of clusters in the input image. This index was proposed by Pakhira *et al* in 2004. The maximum value of *PBM index* indicates optimal result.

Mathematically, PBM index can be defined as follows:

$$PBM = \left(\frac{1}{K} \times fracE_0 E_K \times D_K \right)^2 \tag{5.9}$$

where K represents the number of clusters. The parameter E_0 is constant that can be defined as $E_0 = \sum_{P \in DS} \|P - V\|$ where V represents the center of the patterns $P \in DS$.

The parameter E_K can be defined as $E_K = \sum_{i=1}^{K} \sum_{j=1}^{N} U_{ij} \|P_j - V_i\|$, where N is the number of data points in the data set, $U(DS) = [U_{ij}]_{K \times N}$ is the partition matrix of the data points. V_i represents the center of the ith cluster. D_K signifies the cluster separation measure, which can be defined as $D_K = \max_{i,j=1}^{K} \|V_i - V_j\|$. The details of this cluster validity index are available in [42].

5.5 OVERVIEW OF MANTA RAY FORAGING OPTIMIZATION ALGORITHM

Manta Ray Foraging Optimization algorithm [18] has been proposed by Weiguo Zhao *et al* in 2020, which was inspired by the foraging behaviors of manta rays. Manta rays generally use three unique foraging strategies, viz., chain foraging, cyclone foraging, and somersault foraging. In MRFO, the population has been perturbed by applying these three foraging strategies depending upon some constrains. These strategies have been elaborately described in [18]. The position of the ith manta ray having dimension d at a time stamp t can be represented as $P_i^d(t)$, where $i \in N$.

In chain foraging, Manta rays swim towards the position of plankton by making a foraging chain. During this time, they not only accept the first move towards the food but also they follow their preceding one. The mathematical model used for chain foraging is represented as follows:

$$P_i^d(t+1) = \begin{cases} P_i^d(t) + r.(P_{best}^d(t) - P_i^d(t)) + \alpha.(P_{best}^d(t) - P_i^d(t)), \\ \qquad\qquad i = 1 \\ P_i^d(t) + r.(P_{i-1}^d(t) - P_i^d(t)) + \alpha.(P_{best}^d(t) - P_i^d(t)), \\ \qquad\qquad i = 2, 3, ..., N \end{cases} \tag{5.10}$$

where $\alpha = 2.r.\sqrt{|log(r)|}$ and $r \in [0, 1]$ is a random number.

In cyclone foraging strategy, whenever a shoal of manta rays identifies a patch of plankton in deep water, they make a long chain for foraging and swim towards the food source with a spiral motion. The mathematical model used for cyclone foraging is represented as follows:

$$P_i^d(t+1) = \begin{cases} P_{best}^d + r.(P_{best}^d(t) - P_i^d(t)) + \beta.(P_{best}^d(t) - P_i^d(t)), \\ \qquad\qquad i = 1 \\ P_{best}^d + r.(P_{i-1}^d(t) - P_i^d(t)) + \beta.(P_{best}^d(t) - P_i^d(t)), \\ \qquad\qquad i = 2, 3, ..., N \end{cases} \tag{5.11}$$

where $\beta = 2.e^{r_1(\frac{T-t+1}{T})}.sin(2\pi r_1)$, r_1 is a random number selected between $[0, 1]$ and T represents the maximum number of iteration.

The mathematical model used for this strategy can also be represented as follows:

$$P_i^d(t+1) = \begin{cases} P_{rand}^d + r.(P_{rand}^d(t) - P_i^d(t)) + \beta.(P_{rand}^d(t) - P_i^d(t)), \\ \qquad i = 1 \\ P_{rand}^d + r.(P_{i-1}^d(t) - P_i^d(t)) + \beta.(P_{rand}^d(t) - P_i^d(t)), \\ \qquad i = 2, 3, ..., N \end{cases} \qquad (5.12)$$

It can be noted that, based on some predefined condition, either Equation (5.11) or Equation (5.12) is used in cyclone foraging strategy.

In somersault foraging strategy, each manta ray tries to update its position to the best position found so far by swimming around the food source. It then somersaults to new position. The mathematical model used for somersault foraging is represented as follows.

$$P_i^d(t+1) = P_i^d(t) + S.(r2.P_{best}^d - r3.P_i^d(t)), \quad i = 1, 2, ..., N \qquad (5.13)$$

where the somersault range of manta rays is decided by using somersault factor, $S = 2$. r_2 and r_3 are two random numbers selected from $[0, 1]$.

5.6 PROPOSED METHODOLOGY

In this paper, a novel algorithm has been proposed for *automatic clustering* of color images, named Quantum-Inspired Manta Ray Foraging. In this algorithm, the features of quantum gates, viz., Quantum Rotation and Pauli-X gate, have been used to enhance its computational capability.

In the proposed QIMRFO, the quantum encoding mechanism has been performed to encode each of the *manta ray* of the population for identifying the active cluster centroids from the original population. The basic steps of QIMRFO algorithm are presented as follows:

1. Initially, a population $POP_{[N \times L]}$ with N number of manta rays is created using pixel intensities of a color image. The pixel intensities are chosen randomly from a color image and thereafter these are normalized between 0 to 1 to form the elements of $POP_{[N \times L]}$. In $POP_{[N \times L]}$, each manta ray of length L is considered as a solution of the problem and L is chosen as the square root of the highest intensity value among all the components of pixel intensity (red, green, and blue).

2. The $POP_{[N \times L]}$ is encoded by a quantum state population $QSP_{[N \times L]}$, which contains the quantum states either $|0\rangle$ or $|1\rangle$ depending upon the probability amplitudes of the qubits. The probability amplitudes α_{ij} and β_{ij} of a qubit are produced by randomly generated values of $\theta \in [0, 2\pi]$, which yields the formula as mention in Equation (5.4). The following condition is used to detect the state of a single qubit which belongs to $QSP_{[N \times L]}$.

If $|\alpha_{ij}|^2 < |\beta_{ij}|^2$ then $|\Psi_{ij}\rangle = |0\rangle$, otherwise, $|\Psi_{ij}\rangle = |1\rangle$
where, $i = 1, 2, ..., N$ and $j = 1, 2, ..., L$.

3. The cluster centroids are identified from the original population $POP_{[N\times L]}$ with the help of quantum state population $QSP_{[N\times L]}$. In $QSP_{[N\times L]}$, the values of $|\Psi\rangle$ that are identified as $|1\rangle$, indicate active cluster centroids. Then, fitness FT_N of each individual manta ray is computed by using Equation (5.9).

4. Quantum rotation gate is used to generate the new quantum state population $QSNP_{[N\times L]}$ by using Equation (5.6). The same set of operations as specified in Step 3 are performed again to compute new fitness values NFT_N of each individual *manta ray* by identifying the cluster centroids which belong to $POP_{[N\times L]}$ with the help of $QSNP_{[N\times L]}$.

5. Thereafter, the quantum state population $QSP_{[N\times L]}$ and the fitness FT_N of the manta rays are updated based on the fitness values of FT_N and NFT_N.

6. The basic steps of MRFO are performed on the original population $POP_{[N\times L]}$ to perturb it.

   a. The position of the ith manta ray at a time stamp t is updated by using the following criteria.

      If *rand* < 0.5, then
      i. If $\frac{t}{T_{max}} < rand$ then, Equation (5.12) is used
      ii. Otherwise, Equation (5.11) is used
      Else Equation (5.10) is used.

   b. The fitness FT_N of all manta rays is computed by executing Step 3 and thereafter, the best individual among them is identified.

   c. The positions of all manta rays at time stamp t are updated by Equation (5.13). Thereafter, Step 7 is performed.

7. If the fitness values are not improved, then quantum Pauli-X gate is used (by using Equation (10.24)) based on a predefined mutation probability to achieve diversity in $POP_{[N\times L]}$ and afterward, the Step 3 is performed; otherwise Step 8 is performed.

8. Steps 5 to 9 are repeated for a predefined number of times or until the stopping criteria is met.

9. Finally, the best fitness value and the corresponding number of cluster centroids are reported as the optimal results.

In the proposed QIMRFO, a single manta ray may have more than one fitness values due to the use of quantum rotation gate operation. As the quantum states of that individual are changed due to the operation of quantum rotation gate, there is a possibility that the selection of active cluster centroids will be changed. The execution is always carried out only with the best fitness values of the manta rays, which produces the exploration in the search space. Similarly, the exploitation in the search space is accomplished by the use of *Pauli-X* gate as it acts like a mutation operation which enables the quantum states to flip their values from 0 to 1 and vice versa, depending upon a predefined mutation probability. So, by incorporating the features of quantum gates, the search space can be diversified and it always yields better solution. The flowchart of the proposed algorithm is presented in Figure 5.1.

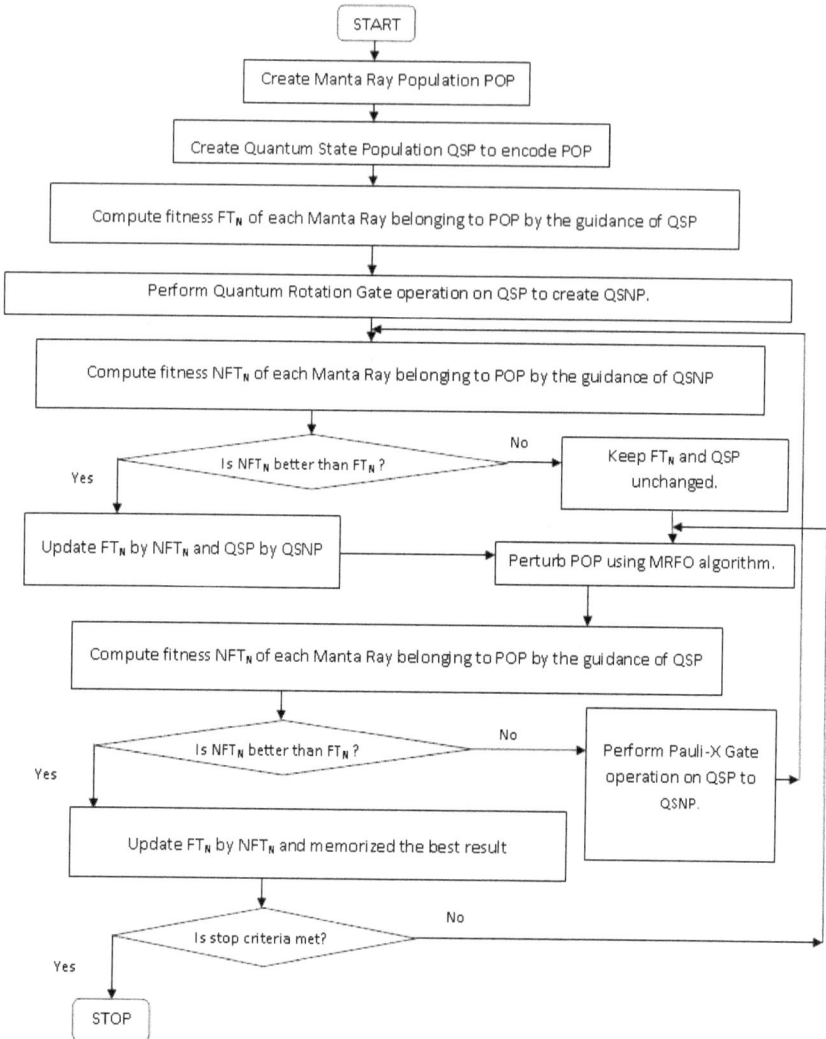

Figure 5.1: Flowchart of QIMRFO algorithm for automatic clustering of color images.

5.7 EXPERIMENTAL RESULTS AND ANALYSIS

This paper presents a Quantum-Inspired Manta Ray Foraging Optimization algorithm to identify the optimum number of clusters in color images on the run. The proposed algorithm has been compared with two different algorithms, viz., MRFO and GA to judge its supremacy. The proposed algorithm uses the quantum mechanical principles, viz., the quantum *rotation* gate and *Pauli-X* gate strategy in its structure. The experimental process and the analysis of the results are presented in the following subsections.

5.7.1 DEVELOPMENTAL ENTERTAINMENT

The experiments have been conducted on four Berkeley [31] and four publicly available real-life color images [44][45]. As a developmental environment, Windows 10 operating system with Intel(R) Core (TM) i5-8250U processor, 8.00GB RAM and 1.60GHz CPU speed, has been used. The Python language has been chosen for implementing the proposed algorithm.

5.7.2 DATASET USED

In this paper, eight color images, four are Berkeley images [31] and four publicly available real-life images [44][45], have been chosen for the experimental purpose. The test images are presented in Figure 5.2, in which (a)-(d) are Berkley images [31] of size 481×321 and (e)-(h) are real life images [44][45] of sizes 510×383, 512×480, 768×512 and 512×480, respectively.

5.7.3 CLUSTERED IMAGES

The clustered images corresponding to the test images are presented in Figure 5.3 and the threshold values obtained by executing the proposed algorithm have been presented in Table 5.4.

5.7.4 SENSITIVITY ANALYSIS OF QIMRFO

The performance of the proposed algorithm has been evaluated by executing it several times with different parameter settings. As the convergence of any algorithm depends upon the proper tuning of parameters, the proposed algorithm has also been gone through Sobol's sensitivity analysis [19][20][21] for its proper tuning. The experiments for MRFO and GA have been conducted with their original parameter values as mentioned in their original works [8][18]. The Python's SALib library has been used to accomplish the sensitivity analysis for identifying the impact of relative contribution of each input parameters toward generating the output.

The conditional variances are evaluated to measure the performance of the analysis. During the evaluation, the *first-order* index (I_{Aj}) is used to measure the direct contribution of each input factor to the output variance, which is represented by the following equation:

$$I_{Aj} = \frac{V\left(\mathbb{EX}[Q \mid P_j]\right)}{V(Q)} \tag{5.14}$$

where P_j represents the jth input parameter and Q represents the output of the model. V represents the variance, and the expectation of Q conditional probability on a particular value of P_j is computed by $\mathbb{EX}[Q \mid P_j]$.

(a)

(b)

(c)

(d)

(e)

(f)

(g)

(h)

Figure 5.2: Data Sets : (a)#22093, (b)#163014, (c)#102061, (d)#159045, (e)#pool, (f)#flower, (g)#tulips and (d)#fruits [31][44][45].

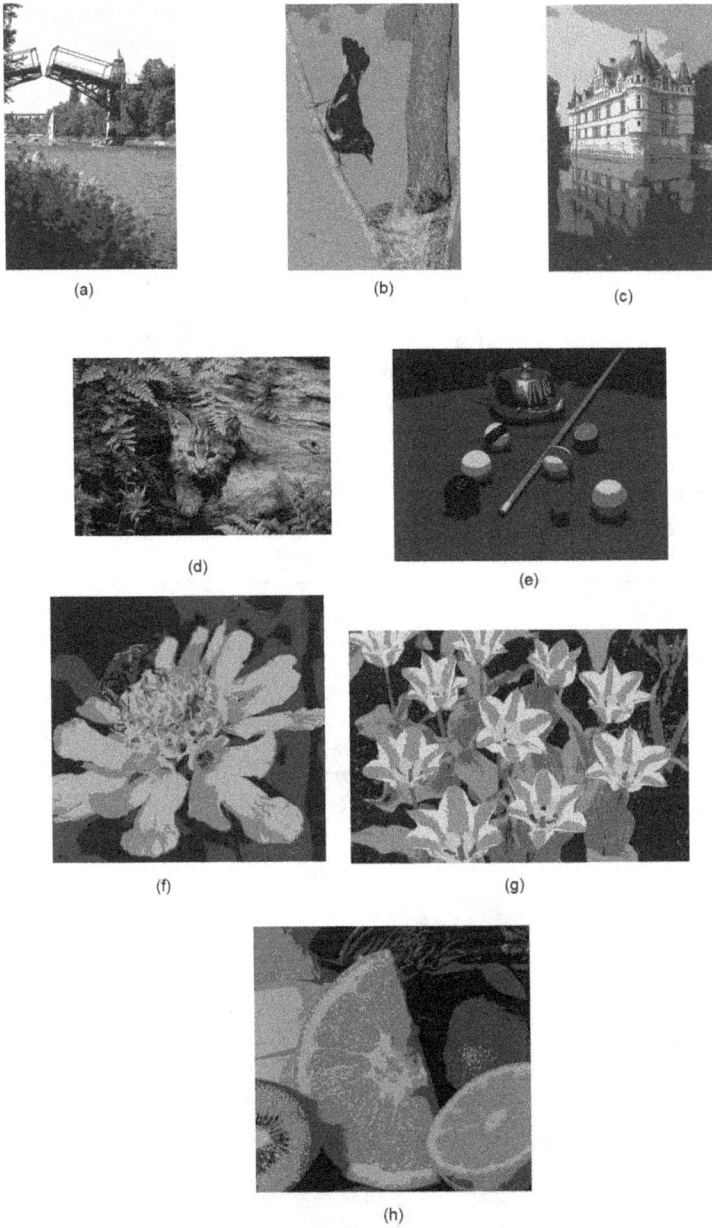

Figure 5.3: Clustered Images of (a)#22093, (b)#163014, (c)#102061, (d)#159045, (e)#pool, (f)#flower, (g)#tulips and (d)#fruits [31][44][45].

Higher values of I_{Aj} as mentioned in Equation (5.14) signify better effectiveness of the output parameter. The measurement of total indices I_{Tj} is used to show the influence of the jth parameter on the output, which can be mathematically represented by

$$I_{Tj} = \frac{\mathbb{EX}[V(Q \mid P_j)]}{V(Q)} \tag{5.15}$$

The output is influenced by the corresponding non-zero value of I_{Tj}.

In this paper, Sobol's *sensitivity analysis* test [19][20][21] has been used to tune four parameters, viz., population size, maximum number of iterations, somersault factor and the mutation probability of the proposed algorithm. The results of the *sensitivity analysis* tests are presented in Table 5.1. The other parameters, viz., α, β, small rotation angle δ as used in QIMRFO are basically random in nature; hence no tuning is required for them. The length of each manta ray has not been tuned since it has been assigned the same value for all the participating algorithms.

After tuning the parameters of QIMRFO, the best possible combination of those parameters is reported in Table 5.2. This table also presents the original settings of parameters of MRFO and GA, which have been mentioned in their respected

---

Table 5.1

Experimental Results of Sensitivity Analysis [19][20][21] for QIMRFO

Parameters	Range	1st Order Effect	Total Effect
Population	10	-1.0714	0.0999
	20	-0.9827	0.1678
	30	**0.7925**	0.5478
	40	0.3803	0.5320
	50	0.2999	0.8432
Maximum Iteration	100	-0.4746	1.0764
	500	0.0632	1.1200
	1000	**0.8305**	1.3829
	1500	0.8211	0.9851
	2000	0.5319	0.8733
Somersault Factor	1.5	0.0461	1.7926
	1.6	0.1297	1.1091
	1.7	**0.4793**	0.8737
	1.8	0.4305	1.2862
	1.9	0.3922	1.7701
	2.0	0.3774	1.6529
Mutation Probability	0.1	-0.6052	0.0542
	0.05	-0.4899	0.0833
	0.03	-0.1701	0.0798
	0.01	**0.6915**	0.0916
	0.005	-0.0762	1.2940
	0.003	0.2824	1.9963
	0.001	0.5350	0.0374

Table 5.2

Settings of Parameters for QIMRFO, MRFO and GA

Parameters	QIMRFO	MRFO	GA
Maximum Iteration : $M_{AX}I$	1000	1000	1000
Population Size: N	30	50	50
Somersault Factor : S_F	1.7	2	-
Crossover Probability : C_P	-	-	0.85
Mutation Probability : M_P	0.01	-	0.001
Small Rotation Angle : δ	[-1.0, 1.0]	-	-

works [8][18]. Each algorithm has been executed for 40 times using these different settings of parameters and the average values among different runs have been considered for reporting purpose.

5.7.5 ANALYSIS OF EXPERIMENTAL RESULTS

In this paper, a novel Quantum-Inspired Manta Ray Foraging Optimization algorithm has been introduced for automatic clustering of color images. The proposed algorithm has been implemented on Python environment. The experiments have been conducted on four Berkeley images [31] and four publicly available real-life color images [44][45] having different dimensions. All the test images are presented in Figure 5.2.

The efficiency of proposed algorithm has been compared with its classical counterpart and the well-known genetic algorithm in different aspects. Comparisons have been made with reference to the mean fitness value, standard deviation of fitness, standard error of fitness, and computational time. The experimental results of all the algorithms have been presented in Tables 5.3–5.6. The optimal number of cluster (η_c), mean fitness value (μ), standard deviation (σ), standard error (ε), and computational time (τ) for all the algorithms are recorded in Table 5.3. The results prove the efficacy of QIMRFO over MRFO and GA. The optimal number of cluster (η_c) and the threshold values of three color components, viz., red, green, and blue, of the proposed algorithm are presented in Table 5.4.

A statistical test, called t-test has also been conducted between the proposed algorithm and each of the comparable algorithm separately. The t-test [38] has been conducted with 95% confidence level to find p-values. If $p < 0.05$, then null hypothesis is rejected and the corresponding alternative hypothesis is accepted. The results of statistical t-test are presented in Table 10.4. While performing t-test between QIMRFO and MRFO, five results are extremely significant and two are quite significant and rest of one is not significant. The test between QIMRFO and GA shows that all eight results are extremely significant.

Finally, with the help of *Friedman test* [47][48], the average ranking has been accomplished for all the algorithms. The test results are presented in Table 5.6. The average rank of QIMRFO, MRFO, and GA has been found to be 1.2, 2.0 and 2.75,

Table 5.3

Number of Cluster (η_c), Mean (μ), Standard Deviation (σ), Standard Error (ε), Computational Time (τ) in Second of QIMRFO, MRFO, and GA

Data Sets	Algorithms	η_c	μ	σ	ε	τ
#22093	QIMRFO	5	**3.369005**	**0.150678**	**0.008699**	**65.97**
	MRFO	5	2.378611	0.294815	0.037547	126.06
	GA	5	2.049843	0.349721	0.060214	162.19
#163014	QIMRFO	4	**0.089985**	**0.001435**	**0.001015**	**73.24**
	MRFO	3	0.077392	0.039875	0.028196	152.54
	GA	4	0.058110	0.047322	0.065918	177.45
#102061	QIMRFO	6	1.836728	**0.173652**	**0.000462**	200.11
	MRFO	6	**1.969722**	0.329838	0.007926	**134.26**
	GA	4	1.373659	0.759327	0.009992	276.08
#159045	QIMRFO	5	**2.060309**	**0.000439**	0.000273	**89.29**
	MRFO	5	1.687840	0.011284	**0.000206**	111.13
	GA	5	1.405063	0.069531	0.009842	151.24
#pool	QIMRFO	4	**22.10002**	**0.059366**	**0.000117**	**112.36**
	MRFO	4	15.36671	0.624183	0.002393	147.10
	GA	4	10.73810	0.547716	0.007555	163,44
#flower	QIMRFO	4	**0.015439**	**0.002204**	**0.000833**	297.28
	MRFO	4	0.009497	0.042827	0.006027	**297.01**
	GA	5	0.006394	0.007843	0.005196	312.14
#tulips	QIMRFO	5	**0.418231**	0.000438	0.002458	**356.29**
	MRFO	5	0.292812	**0.000126**	**0.000972**	382.15
	GA	5	0.248261	0.004701	0.067391	430.38
#fruits	QIMRFO	4	**2.575446**	**0.000278**	**0.000197**	**86.45**
	MRFO	4	1.528733	0.023288	0.016467	191.62
	GA	4	1.793581	0.019843	0.008652	222.47

respectively. The test values prove that QIMRFO is the best performing algorithm among all of the competitive algorithms.

The convergence curves of QIMRFO, MRFO, and GA are shown in Figure 5.4. Each figure shows that QIMRFO converges faster than others. Hence, the superiority of the proposed algorithm has been visually and quantitatively established using different measures. The population diversity curves using the quantum *rotation* gate and the *Pauli-X* gate have been presented in Figures 5.5 and 5.6, respectively.

5.8 CONCLUSION AND FUTURE SCOPE

In this paper, a Quantum-Inspired Manta Ray Foraging Optimization algorithm has been presented for automatic clustering of color images. A novel quantum rotation gate and Pauli-X gate strategy have been introduced to achieve the exploration and

Table 5.4

Number of Cluster (η_c) with its Corresponding Threshold Values of the Clustered Images

Data Sets	η_c	Color Component	Threshold Value
#22093	5	R	[30, 86, 100, 135, 232]
		G	[51, 92, 143, 175, 247]
		B	[36, 75, 140, 169, 188]
#163014	4	R	[40, 103, 140, 189]
		G	[42, 125, 150, 176]
		B	[20, 75, 100, 153]
#102061	6	R	[47, 80, 91, 132, 145, 240]
		G	[62, 100, 155, 210, 229, 241]
		B	[51, 74, 135, 212, 240, 250]
#159045	5	R	[39, 75, 112, 147, 180]
		G	[30, 61, 100, 143, 175]
		B	[11, 25, 63, 95, 140]
#pool	4	R	[15, 25, 140, 200]
		G	[43, 61, 90, 192]
		B	[10, 53, 142, 191]
#flower	4	R	[50, 85, 158, 200]
		G	[25, 60, 106, 153]
		B	[10, 42, 147, 196]
#tulips	5	R	[25, 61, 83, 212, 229]
		G	[30, 100, 112, 146, 200]
		B	[31, 62, 80, 150, 191]
#fruits	4	R	[37, 77, 148, 162]
		G	[10, 62, 100, 146]
		B	[25, 36, 70, 98]

exploitation in the search space for identifying the optimal results. Several tests have been conducted among the competitive algorithms to judge the effectiveness of the proposed algorithm. The experimental results prove that QIMRFO outperforms others in all aspects.

In future, the functionality of the proposed algorithm can be expanded in such a way that it can efficiently handle the high dimensional data sets. Various quantum gate strategies can also be used to develop new algorithms for solving different types of optimization problems.

Table 5.5

Results of t-Test Between QIMRFO vs. MRFO, and QIMRFO vs. GA

Data Sets		p - value	Significance
#22093	QIMRFO & MRFO	<0.0001	Extremely Significant
	QIMRFO & GA	<0.0001	Extremely Significant
#163014	QIMRFO & MRFO	0.04571	Significant
	QIMRFO & GA	<0.0001	Extremely Significant
#102061	QIMRFO & MRFO	0.0268	Significant
	QIMRFO & GA	0.0003	Extremely Significant
#159045	QIMRFO & MRFO	<0.0001	Extremely Significant
	QIMRFO & GA	<0.0001	Extremely Significant
#pool	QIMRFO & MRFO	<0.0001	Extremely Significant
	QIMRFO & GA	<0.0001	Extremely Significant
#flower	QIMRFO & MRFO	0.3835	Not Significant
	QIMRFO & GA	<0.0001	Extremely Significant
#tulips	QIMRFO & MRFO	<0.0001	Extremely Significant
	QIMRFO & GA	<0.0001	Extremely Significant
#fruits	QIMRFO & MRFO	<0.0001	Extremely Significant
	QIMRFO & GA	<0.0001	Extremely Significant

Table 5.6

Results of Friedman Test [47][48] for QIMRFO, MRFO and GA

Data Sets	QIMRFO	MRFO	GA
#22093	3.9218 (1)	2.7263 (2)	2.4497 (3)
#163014	0.0921 (1.5)	0.0921 (1.5)	0.0765 (3)
#102061	1.9728 (2)	2.1097 (1)	1.7383 (3)
#159045	2.0999 (1)	1.6982 (2.5)	1.6982 (2.5)
#pool	22.8993 (1)	15.9609 (2)	14.9761 (3)
#flower	0.0171 (1.5)	0.0171 (1.5)	0.0162 (3)
#tulips	0.4314 (1)	0.2983 (2.5)	0.2983 (2.5)
#fruits	2.5833 (1)	1.5934 (3)	1.8217 (2)
Average Rank :	1.25	2.0	2.75

ACKNOWLEDGMENT

This work was supported by the AICTE sponsored RPS project on Automatic Clustering of Satellite Imagery using Quantum-Inspired Metaheuristics vide F.No 8-42/RIFD/RPS/Policy-1/2017-18.

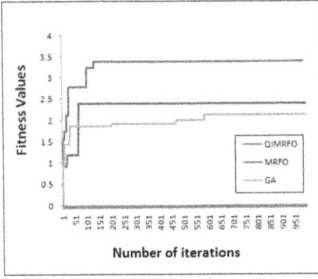

(a) #22093

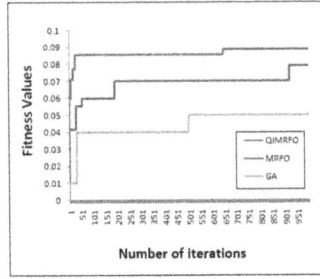

(b) #163014

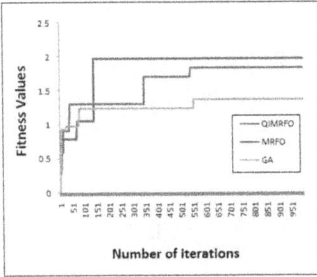

(c) #102061

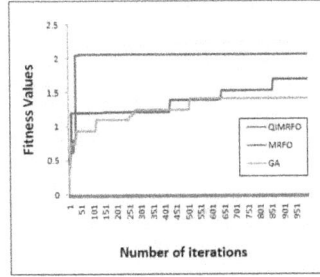

(d) #159045

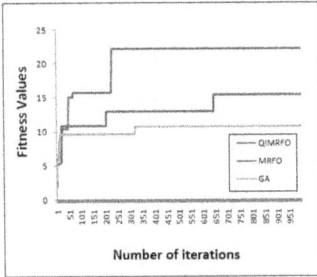

(e) #pool

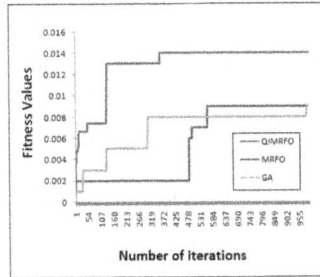

(f) #flower

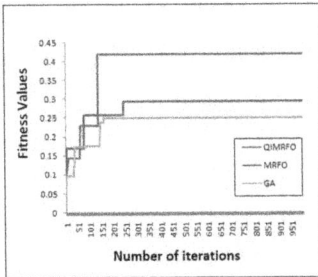

(g) #tulips

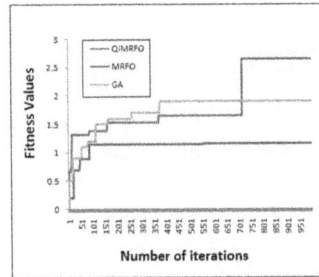

(h) #fruits

Figure 5.4: Convergence curves of QIMRFO, MRFO, and GA for test images.

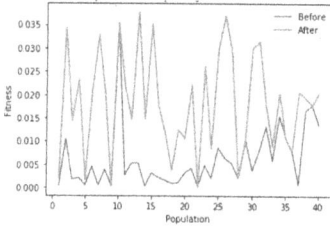

(a) #22093

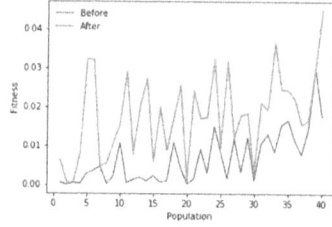

(b) #163014

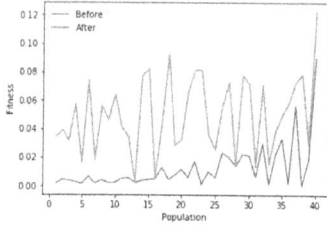

(c) #102061

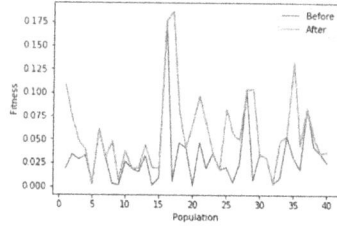

(d) #159045

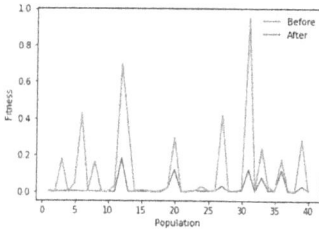

(e) #pool

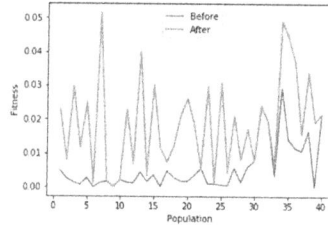

(f) #flower

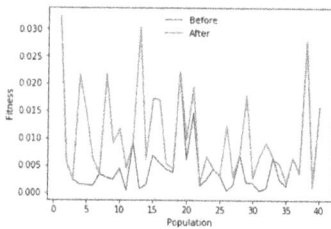

(g) #tulips

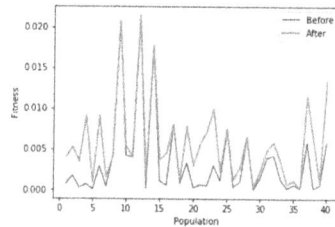

(h) #fruits

Figure 5.5: Population Diversity using Quantum Rotation Gate of test images [31][44][45].

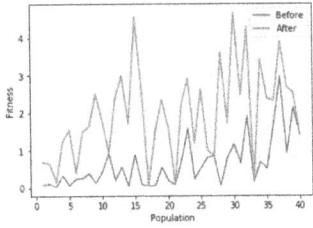

(a) #22093

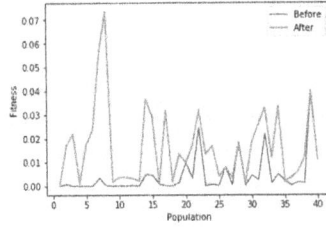

(b) #163014

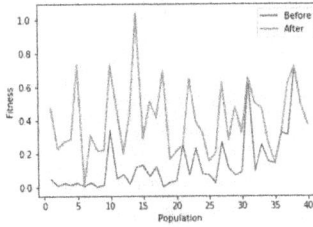

(c) #102061

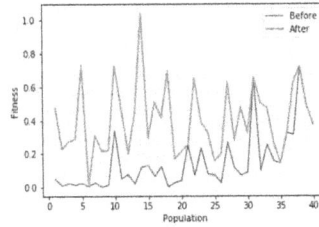

(d) #159045

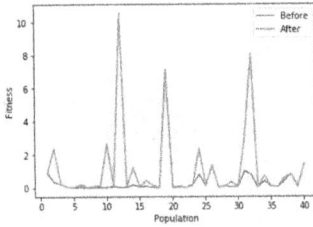

(e) #pool

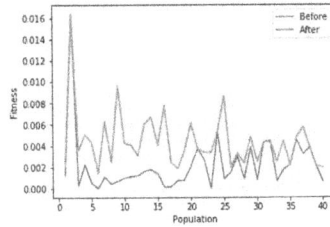

(f) #flower

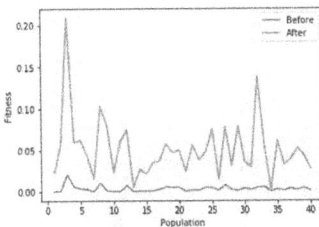

(g) #tulips

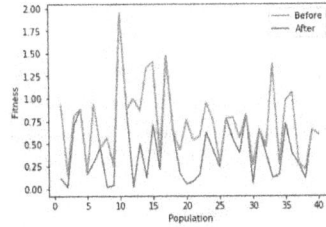

(h) #fruits

Figure 5.6: Population Diversity Increased by using Pauli-X Gate of test images [31][44][45].

REFERENCES

1. A. K. Jain and R. C. Dubes. Algorithms for Clustering Data. Prentice-Hall, Inc., USA, 1988.

2. A. K. Jain, M. N. Murty, and P. J. Flynn. Data clustering: A review. ACM Computing Surveys, 31(3):264–323, 1999.

3. S. Bandyopadhyay and U. Maulik. Genetic clustering for automatic evolution of clusters and application to image classification. Pattern Recognition, 35(6):1197–1208, 2002.

4. A. E. Ezugwu. Nature-inspired metaheuristic techniques for automatic clustering: A survey and performance study. SN Applied Sciences, 2, 2020.

5. A. Jose-Garca and W. Gomez-Flores. Automatic clustering using nature-inspired metaheuristics: A survey. Applied Soft Computing, 41:192–213, 2016.

6. S. Das, A. Abraham, and A. Konar. Automatic clustering using an improved differential evolution algorithm. IEEE Transactions on Systems, Man, and Cybernetics – Part A: Systems and Humans, 38(1):218–237, 2008.

7. John H. Holland. Adaptation in Natural and Artificial Systems: An Introductory Analysis with Applications to Biology, Control and Artificial Intelligence. MIT Press, Cambridge, MA, 1992.

8. J. Kennedy and R. Eberhart. Particle swarm optimization. In Proc. IEEE International Conference on Neural Networks, Perth, Australia, pp. 1942–1948, 1995.

9. R. Storn and K. Price. Differential evolution – a simple and efficient heuristic for global optimization over continuous spaces. Journal of Global Optimization, 11:341–359, 1997.

10. S. Bhattacharyya, V. Snasel, A. Dey, S. Dey, and D. Konar. Quantum spider monkey optimization (QSMO) algorithm for automatic gray-scale image clustering. International Conference on Advances in Computing, Communications and Informatics (ICACCI 2018), pp. 1869–1874, 2018.

11. A. Dey, S. Bhattacharyya, S. Dey, J. Platos, and V. Snasel. Quantum-inspired bat optimization algorithm for automatic clustering of grayscale images, vol. 922, pp. 89–101. Springer, Singapore, 2019.

12. A. Dey, S. Dey, S. Bhattacharyya, J. Platos, and V. Snasel. Novel quantum-inspired approaches for automatic clustering of gray level images using particle swarm optimization, spider monkey optimization and ageist spider monkey optimization algorithms. Applied Soft Computing, 88(106040), 2020.

13. A. Dey, S. Dey, S. Bhattacharyya, J. Platos, and V.S. Snasel. Quantum-Inspired Automatic Clustering Algorithms: A comparative study of Genetic Algorithm and Bat Algorithm, pp. 89–114. De Gruyter, 2020.

14. A. Dey, S. Dey, S. Bhattacharyya, V. Snasel, and A.E. Hassanien. Simulated Annealing Based Quantum-Inspired Automatic Clustering Technique, pp. 73–81. Cairo, 2018.

15. S. Dey, S. Bhattacharyya, and M. Ujjwal. Quantum Behaved Swarm Intelligent Techniques for Image Analysis: A Detailed Survey, pp. 1–39. GI Global, Hershey, USA, 2015.

16. S. Dey and U. Bhattacharyya and S. Maulik. Quantum-Inspired Automatic Clustering Technique Using Ant Colony Optimization Algorithm, 2018.

17. K.H. Han and J.H. Kim. Quantum-inspired evolutionary algorithm for a class of combinatorial optimization. IEEE Transactions on Evolutionary Computation, 6(6):580–593, 2002.

18. W. Zhao, Z. Zhang, and L. Wang. Manta ray foraging optimization: An effective bio-inspired optimizer for engineering applications. Engineering Applications of Artificial Intelligence, 87:103300, 2020.

19. A. Saltelli, P. Annoni, I. Azzini, F. Campolongo, M. Ratto, and S. Tarantola. Variance based sensitivity analysis of model output. design and estimator for the total sensitivity index. Computer Physics Communications, 181(2):259–270, 2010.

20. A. Saltelli and I.M. Sobol. Sensitivity analysis for nonlinear mathematical models: Numerical experience. Matematicheskoe Modelirovanie, 7(11):16–28, 1995.

21. I.M. Sobol. Global sensitivity indices for nonlinear mathematical models and their Monte Carlo estimates. Mathematics and Computers in Simulation, 55(1–3):271–280, 2001.

22. H. Frigui and R. Krishnapuram. A robust competitive clustering algorithm with applications in computer vision. IEEE Transactions on Pattern Analysis and Machine Intelligence, 21(5):450–465, 1999.

23. P.S. Bradley and U.M. Fayyad. Refining initial points for k-means clustering. In Proceedings of the Fifteenth International Conference on Machine Learning, pp. 91–99. Morgan Kaufmann Publishers Inc., 1998.

24. D. Pelleg and A. Moore. X-means: Extending k-means with efficient estimation of the number of clusters. In Proceedings of the 17th International Conference on Machine Learning, pp. 727–734. Morgan Kaufmann, 2000.

25. X.L. Meng and D.V. Dyk. The em algorithm an old folksong sung to a fast new tune. Journal of the Royal Statistical Society. Series B (Methodological), 59(3):511–567, 1997.

26. F. Murtagh. A survey of recent advances in hierarchical clustering algorithms. The Computer Journal, 26(4):354–359, 1983.

27. F.J. Rohlf. 12 single-link clustering algorithms. In Classification Pattern Recognition and Reduction of Dimensionality, vol. 2 of Handbook of Statistics, pp. 267–284. Elsevier, 1982.

28. M. Dorigo, M. Birattari, and T. Stutzle. Ant colony optimization, vol. 1. IEEE, 2006.

29. X.S. Yang. A new metaheuristic bat-inspired algorithm. In Nature Inspired Cooperative Strategies for Optimization (NICSO 2010), pp. 65–74. Springer, 2010.

30. K.M. Passino. Biomimicry of bacterial foraging for distributed optimization and control. IEEE Control Systems Magazine, 22(3):52–67, 2002.

31. X.-S. Yang. Firefly algorithm, stochastic test functions and design optimisation. International Journal of Bio-inspired Computation, 2(2):78–84, 2010.

32. X.-S. Yang and S. Deb. Cuckoo search via levy flights. pp. 210–214, 2010.

33. A Askarzadeh. A novel metaheuristic method for solving constrained engineering optimization problems: Crow search algorithm. Computers & Structures, 169:1–12, 2016.

34. Z.A. Babak, B.H. Omid, and X. Chu. Crow Search Algorithm (CSA), pp. 143–149. Springer Singapore, Singapore, 2018.

35. T. Gandhi, Nitin, and T. Alam. Quantum genetic algorithm with rotation angle refinement for dependent task scheduling on distributed systems. In 2017 Tenth International Conference on Contemporary Computing (IC3), pp. 1–5. IEEE, Aug 2017.

36. H.P. Chiang, Y.H. Chou, C.H. Chiu, S.Y. Kuo, and Y.M. Huang. A quantum-inspired tabu search algorithm for solving combinatorial optimization problems. Soft Computing, 18:1771–1781, 2013.

37. M. Mahseur, A. Ramdane-Cherif, D. Acheli, and Y. Meraihi. A quantum-inspired binary firefly algorithm for qos multicast routing. International Journal of Metaheuristics, 6(4):309, 2017.

38. S. Dey, S. Bhattacharyya, and U. Maulik. Efficient quantum inspired metaheuristics for multi-level true color image thresholding. Applied Soft Computing, 56:472–513, 2017.

39. S. Dey, I. Saha, S. Bhattacharyya, and U. Maulik. Multilevel thresholding using quantum-inspired metaheuristics. Knowledge-Based System, 67:373–400, 2014.

40. S.S. Tirumala. A quantum-inspired evolutionary algorithm using Gaussian distribution-based quantization. Arabian Journal for Science and Engineering, 43:471–482, 2018.

41. Y.-J. Yang, S.Y. Kuo, F.-J. Lin, I.-I. Liu, and Y.-H. Chou. Improved quantum-inspired tabu search algorithm for solving function optimization problem. 2013 IEEE International Conference on Systems, Man, and Cybernetics, pp. 823–828, 2013.

42. M. Pakhira, S. Bandyopadhyay, and M. Ujjwal. Validity index for crisp and fuzzy clusters. Pattern Recognition, 37:487–501, 2004.

43. Berkley images. www2.eecs.berkeley.edu/Research/ Projects/CS/vision/bsds/BSDS300/ html/dataset/ images.html. Accessed on 15/01/2020.

44. Real life images. www.hlevkin.com/06testimages.htm. Accessed on 15/01/2020.

45. Real life images. https://homepages.cae.wisc.edu/ ece533/images/. Accessed on 23/05/2020.

46. B. Flury. A First Course in Multivariate Statistics. Springer Texts in Statistics

47. M. Friedman. The use of ranks to avoid the assumption of normality implicit in the analysis of variance. Journal of the American Statistical Association, 32(200):675–701, 1937.

48. M. Friedman. A comparison of alternative tests of significance for the problem of m rankings. Annals of Mathematical Statistics, 11(1):86– 92, 1940.

6 Automatic Feature Selection for Coronary Stenosis Detection in X-Ray Angiograms Using Quantum Genetic Algorithm

6.1 INTRODUCTION

The automatic coronary stenosis detection problem is a challenging task since it involves detailed analysis over X-Ray Coronary Angiograms (XCA) in the form of gray-level digital images. XCA remains as the gold-standard imaging technique for medical diagnosis of arterial diseases, including stenosis and other related conditions. In this procedure, a liquid dye, such as fluorescein, is injected using a thin catheter inserted into an access point to the bloodstream (usually in arm or groin). The dye reveals an arterial structure that can be easily seen on X-ray images and allows to cardiologists to detect narrowed or blocked areas through the coronary arteries.

In clinical practice, the stenosis cases detection process is performed by cardiologists. To detect the possible stenosis cases, the specialist performs a visual scan over an X-ray coronary angiogram (XCA) image which can be printed over a physical media or as gray-level digital image. Until the process, the specialist labels the angiogram over different regions where a stenosis case is present according with his expertise and knowledge. Figure 6.1 illustrates an angiogram and their respective stenosis regions labeled by a specialist. However, given the limited access to such delicate clinical expertise, the variability of diagnoses among specialists has allowed that the automatic Computer-Aided Diagnosis (CAD) systems play a vital role in cardiology to assist the detection of coronary artery stenosis.

The problem of automated stenosis cases detection in XCA has been addressed from different approaches in literature. For example, Kishore and Jayanthi [1] make use of a fixed-size window (patch) that is manually selected from an previously enhanced image and after, an adaptive threshold algorithm is applied to keep only vessel pixels. With the segmented image, a vessel width is calculated by adding the intensity values from the left to the right edge. In this approach, there is no need to a

DOI: 10.1201/9781003283294-6

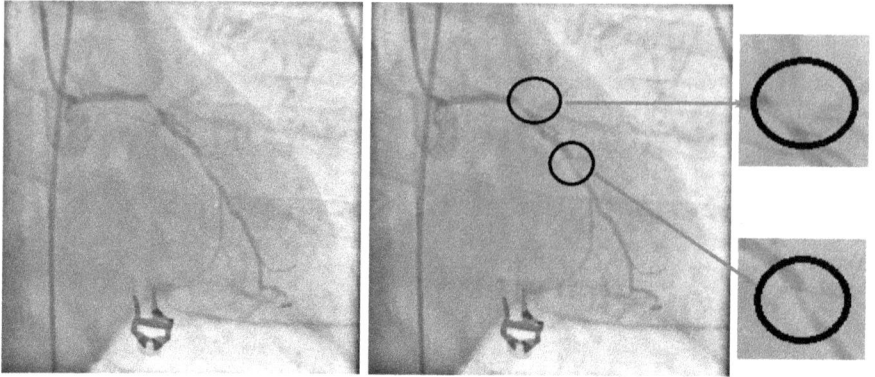

Figure 6.1: X-ray coronary angiogram. From left to right: Original angiogram, stenosis diseases labeled by a cardiologist and, the zoom of the stenosis labels.

skeletonization process of the arteries, using only the vessel width to determine the grade of a stenosis case into the selected window. Saad [2] was able to detect stenosis cases by using vessel skeletons. This approach requires a previous image segmentation process in order to extract vessel pixels and after, a skeletonization procedure is performed to extract only the center lines corresponding to the vessels. With the vessel center lines (skeleton), the length of the orthogonal line is computed using a fixed-size window that is moving over the image in order to obtain a vessel-width measure that is compared with a fixed value to determine if a stenosis case exists or not in that region of the image. Sameed et al. [3] make use of the Hessian matrix to enhance vessel pixels and determine candidate stenosis regions, by identifying narrowed vessel areas. Wan et al. [4] carried out the vessel diameter estimation using a smoothed vessel centerline curve from the candidate stenosis regions detected by the Hessian-approach. Both, from the artery lumen (vessel diameter), allow determining the stenosis measurement and final classification. Cervantes-Sanchez et al. [5] proposed a method for computing the vessel width along the arteries by applying the second-order derivatives directly over the enhanced images, where the cases of stenosis were labeled as local minima through the vessel width; in this approach, no additional skeletonization or vessel diameter estimation was needed. Posteriorly, Cruz-Aceves et al. [6] used a Bayes classifier over a handcrafted 3D feature vector that was obtained from the results of potential cases of coronary stenosis identified previously by a second-order derivative operator. Major disadvantage for this method relies in the need of a predefined threshold or fixed value in the form of a vessel width, narrowest measure, etc., which a current computed value, must be compared or contrasted against to determine if it is greater, equal or lower than that fixed threshold, resulting in a problem by itself the determination of that fixed value.

Convolutional Neural Networks (CNNs) have emerged to overcome the disadvantages of methods where *a-priori* fixed value (or a set of them) must be established to perform a classification. They have been successfully used to solve diverse kinds of problems involving image processing in medical imaging, such as image segmentation, enhancement, classification, and location of specific interest regions. Instead of dealing with pre-defined features, CNN's input are raw images, which are segmented although the convolutional layers by applying a set of specific filters in order to keep only useful information according to the expected results during the training stage. One of the most reliable CNN architectures in medical imaging is the U-Net [7] proposed by Harouni in 2018. U-Net is composed of 18 convolutional layers plus the output layer. Figure 6.2 illustrates the U-Net architecture. In addition,

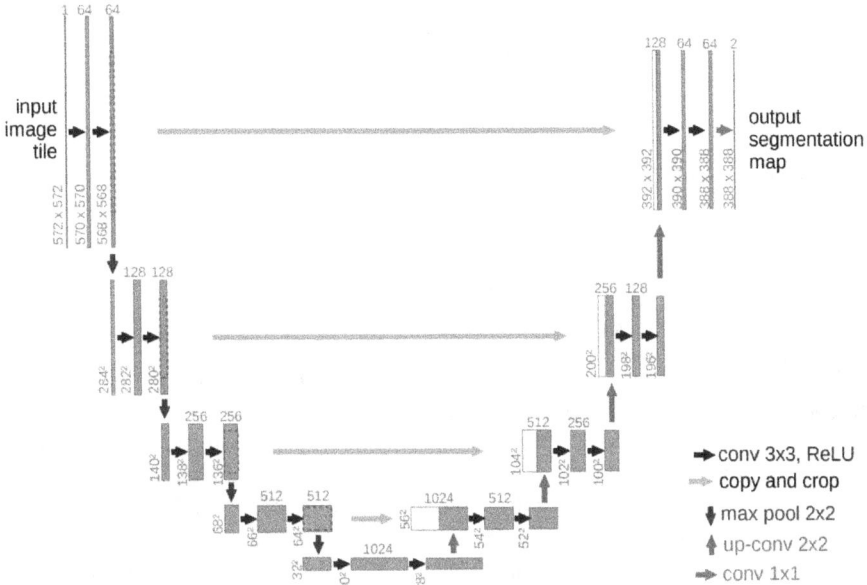

Figure 6.2: U-Net architecture.

CNNs have been applied successfully for the coronary stenosis detection problem. Antczak et al. [8] tested several CNNs architectures using a natural image dataset plus 10000 instances of synthetically generated vessels patches, which includes positive and negative cases. An additional strength of CNNs is the possibility to pass knowledge from a pre-trained CNN to a new one. This process is called *Transfer Learning* [9]. However, success of knowledge transfer depends on dissimilarity between the source domain (where the CNN has been trained) and the target domain (where the knowledge is transferred) [10].

Major disadvantages of CNNs are related with choosing the right CNN architecture and the large amount of instances required to achieve a correct training and classification, despite the risk to fall in an over-fitting state due mainly to an unbalanced dataset (where the number of positive and negative cases are significantly different). For instance, in the automated coronary stenosis detection problem, the number of positive cases are significantly smaller than negative cases. To overcome this

disadvantage, strategies such as data augmentation [11] and synthetic data generation [12] are applied. However, those strategies entails the inconvenient of manually
select a representative and significant dataset from which the new augmented data
will be generated or a correct methodology or model to generate synthetic cases
closed to the real ones.

In this research is proposed a novel method based on the automated feature selection for detection of coronary stenosis cases in X-ray coronary angiograms. The
method was tested using an image database with ≈ 2800 instances. In addition, the
proposed method was compared with other five strategies taken from the state-of-
the-art, which includes machine learning and deep learning techniques. To measure
the effectiveness of the proposed and compared methods, the Accuracy and the Jaccard's Index metrics were used, achieving a rate of 0.92 and 0.85, for the accuracy
and Jaccard index, respectively. The obtained results probed the method effectiveness
for the automated stenosis cases detection problem in X-ray coronary angiograms by
selecting only a subset of features and keeping at same time, an optimal classification
rate.

6.2 BACKGROUND

6.2.1 FEATURE EXTRACTION

In Image Processing, feature extraction term refers to the multiple metrics that can
be computed or *extracted* from an image, a region, or a single pixel from it. Since
single-pixel metrics does not offer valuable information by itself, the most common
approach to extract significant information from an image are the windows (patches)
or the entire image. The feature extraction process can be performed in an automated
way, for instance, by using a CNN. However, almost all CNN-based procedures ob-
fuscate the extracted features because the complexity inherent to the process or the
feature itself. This kind of strategies leads to produce a classification result only.
On the other hand, specific or manually extracted features can be analyzed in detail.
However, this approach has the disadvantage that features must be collected by the
researcher in a non-fully automated way most of the time.

From an overall point of view, extracted features from an image can be classified
in three categories: intensity, texture, and morphology. Each of these groups will
be described on next sections and how they are related with the studied problem.
It is important to mention that, since the gray-level scale pixel intensity is the most
common approach for X-ray coronary angiogram digital imaging, all features will
be related to work with gray-level images.

6.2.1.1 Pixel Intensity-based Features

Pixel intensity features are the most single metrics that can be computed from an image or region of it. They are extracted computing classical statistical measurements
from the intensities (values) of the pixels such as minimum (min), maximum (max),
mean, median, and standard deviation. Figure 6.3 illustrates two images with their
respective intensity values and their corresponding intensity-based feature values.

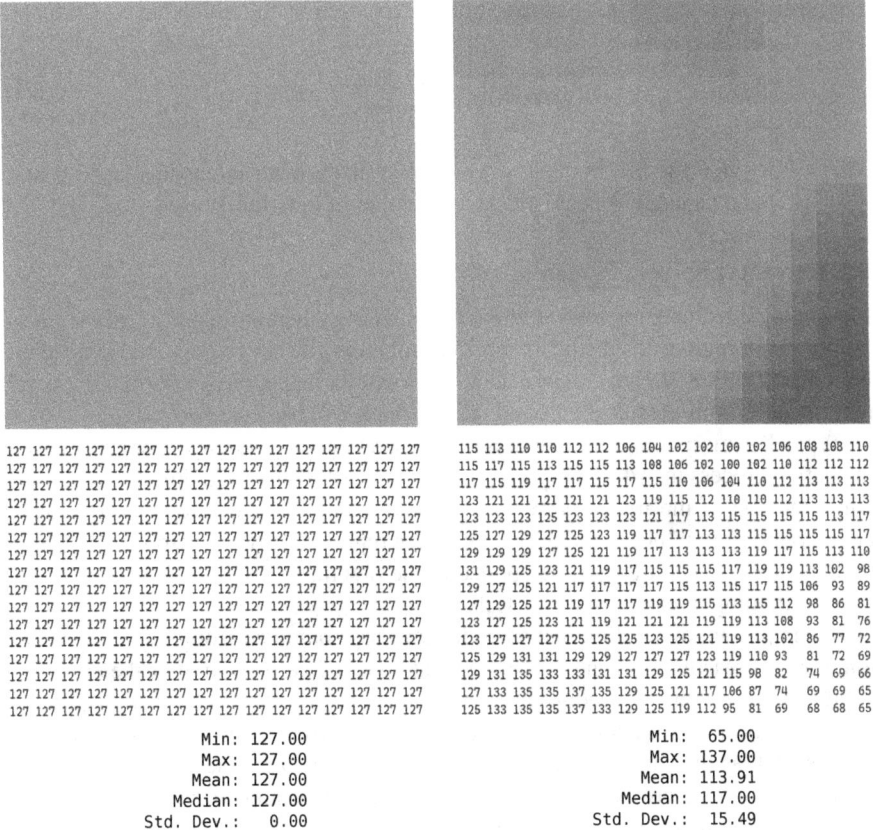

```
127 127 127 127 127 127 127 127 127 127 127 127 127 127 127 127      115 113 110 110 112 112 106 104 102 102 100 102 106 108 108 110
127 127 127 127 127 127 127 127 127 127 127 127 127 127 127 127      115 117 115 113 115 115 113 108 106 102 100 102 110 112 112 112
127 127 127 127 127 127 127 127 127 127 127 127 127 127 127 127      117 115 119 117 117 115 117 115 110 106 104 110 112 113 113 113
127 127 127 127 127 127 127 127 127 127 127 127 127 127 127 127      123 121 121 121 121 121 123 119 115 112 110 110 112 113 113 113
127 127 127 127 127 127 127 127 127 127 127 127 127 127 127 127      123 123 123 125 123 123 123 121 117 113 115 115 115 115 113 117
127 127 127 127 127 127 127 127 127 127 127 127 127 127 127 127      125 127 129 127 125 123 119 117 117 113 113 115 115 115 115 117
127 127 127 127 127 127 127 127 127 127 127 127 127 127 127 127      129 129 129 127 125 121 119 117 113 113 113 119 117 115 113 110
127 127 127 127 127 127 127 127 127 127 127 127 127 127 127 127      131 129 125 123 121 119 117 115 115 115 117 119 119 113 102  98
127 127 127 127 127 127 127 127 127 127 127 127 127 127 127 127      129 127 125 121 117 117 117 117 115 113 115 117 115 106  93  89
127 127 127 127 127 127 127 127 127 127 127 127 127 127 127 127      127 129 125 121 119 117 117 119 119 115 113 115 112  98  86  81
127 127 127 127 127 127 127 127 127 127 127 127 127 127 127 127      123 127 125 123 121 119 121 121 121 119 119 113 108  93  81  76
127 127 127 127 127 127 127 127 127 127 127 127 127 127 127 127      123 127 127 127 125 125 125 125 123 125 121 119 113 102  86  77  72
127 127 127 127 127 127 127 127 127 127 127 127 127 127 127 127      125 129 131 131 129 129 127 127 127 123 119 110  93  81  72  69
127 127 127 127 127 127 127 127 127 127 127 127 127 127 127 127      129 131 135 133 133 131 131 129 125 121 115  98  82  74  69  66
127 127 127 127 127 127 127 127 127 127 127 127 127 127 127 127      127 133 135 135 137 135 129 125 121 117 106  87  74  69  69  65
127 127 127 127 127 127 127 127 127 127 127 127 127 127 127 127      125 133 135 135 137 133 129 125 119 112  95  81  69  68  68  65
```

Min:	127.00	Min:	65.00
Max:	127.00	Max:	137.00
Mean:	127.00	Mean:	113.91
Median:	127.00	Median:	117.00
Std. Dev.:	0.00	Std. Dev.:	15.49

Figure 6.3: Intensity-based features for two images of 16x16 in the first row. Second row contains the corresponding pixel intensities for each of them. Third row contains the corresponding intensity-basic features.

6.2.1.2 Texture Features

Texture features provide information about the *rugosity* level present in an image. Normally, texture features are related with sudden intensity variations in a region of the image. Intensity changes are useful to describe interest objects or regions inside the image. Most common approach to extract texture related features are the Haralick texture features [13], which are computed using the Gray-Level Co-Occurrence Matrix or GLCM. The GLCM is a matrix in which rows and columns are formed by the pixel intensities taken from the entire image or a region from it. The GLCM has the same number of rows and columns. The matrix contents are computed based on the frequency of a pixel with intensity *i* occurring in a specific spatial relationship to a pixel with intensity j. This operation is expressed by the following equation:

$$P(i,j,d,\theta) = \sum_{i=0}^{G-1}\sum_{j=0}^{G-1} P(i,j), \qquad (6.1)$$

where $P(i,j,d,\theta)$ is the frequency which two pixels with intensities i and j at a distance d and an angle θ occurs. G is the number of gray levels used.

6.2.1.3 Morphologic Features

Morphologic features are related with shape and the inherent properties of the interest objects or regions inside the image. For problems related to artery study, morphologic features provide information about the vessel's shape, the proportion of vessel pixels against non-vessels pixels and others which will be described later.

Typically, morphologic features require a previous image enhancement or segmentation process in order to highlight or keep only the interested objects or regions, while the rest of the image is treated as background. Most of the coronary X-ray angiograms contain noise and narrowed artery regions. For this reason, an effective segmentation procedure must be applied to the raw image in order to be able to extract morphologic features from vessels.

It is important to mention that some features are computed from the vessel skeleton. For this reason, a second processing task must be performed after the enhancement or segmentation procedure in order to extract the vessel's center lines that are known as the artery skeleton. The artery skeleton provides information about the vessels such as their tortuosity, number of branches, their segment length, etc. Figure 6.4 illustrates a raw X-ray coronary angiogram image and their respective vessel segmentation and skeleton extraction.

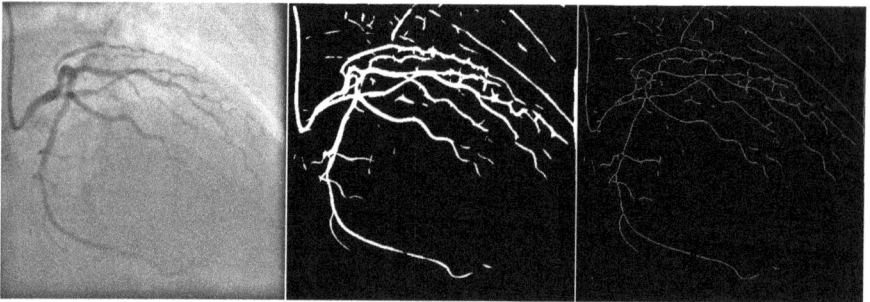

Figure 6.4: Sample of a raw X-ray coronary angiogram image and their corresponding vessel segmentation and skeleton extraction, from left to right.

Additional texture and morphologic features can be extracted by combining information from the raw image and their corresponding vessels and skeleton. For instance, vessel texture measures can be extracted with the segmented vessel information from the raw image.

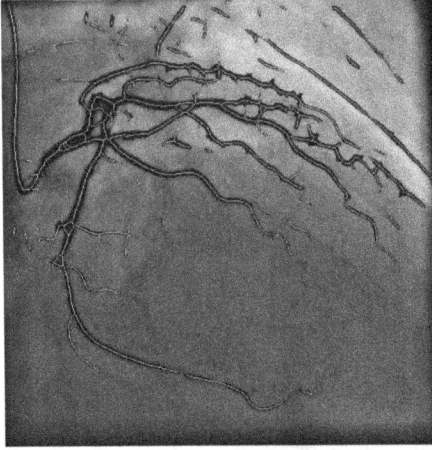

Figure 6.5: The same above X-ray angiogram with their respective vessel segmentation and skeleton overlapped in order to illustrate a combined feature extraction.

For this study, the Frangi [14] method was used in order to segment vessels from the angiographies because the optimal results are obtained. The Frangi method uses the eigenvalues obtained from a Hessian matrix. The Hessian matrix is computed from the second-order derivative of the original image. It is calculated by convolving a Gaussian kernel at different orientations with the original image as follows:

$$G(x,y) = -\exp\left(-\frac{x^2+y^2}{2\sigma^2}\right) \|y\| < L/2, \tag{6.2}$$

where σ is the spread of the Gaussian profile and L is the length of the vessel segment. The resultant Hessian matrix is expressed as follows:

$$H = \begin{pmatrix} H_{xx} & H_{xy} \\ H_{yx} & H_{yy} \end{pmatrix}, \tag{6.3}$$

where H_{xx}, H_{xy}, H_{yx}, and H_{yy} are the directional second-order partial derivatives of the image.

The segmentation function defined by Frangi for 2-D vessel detection is as follows:

$$f(x) = \begin{cases} 0 & \text{if } \lambda_2 > 0, \\ \exp\left(-\frac{R_b^2}{2\alpha^2}\right)\left(1 - \exp\left(\frac{S^2}{2\beta^2}\right)\right) & \text{elsewhere.} \end{cases} \tag{6.4}$$

The α parameter is used with R_b to control the shape discrimination. The β parameter is used by S^2 for noise elimination. R_b and S^2 are calculated as follows:

$$R_b = \frac{|\lambda_1|}{|\lambda_2|}, \tag{6.5}$$

$$S^2 = \sqrt{\lambda_1^2 + \lambda_2^2}, \tag{6.6}$$

where λ_1 and λ_2 are the eigenvalues of Hessian matrix.

The Frangi method response is a gray-scale image with the vessel pixels enhanced. In order to fully eliminate non-vessel pixels (background and noise), a binarization of the Frangi response must be applied. In this research, the Otsu [15] method was used because the threshold value is calculated automatically based on the image pixels by computing the weighted sum of variance of the two classes, expressed as follows:

$$\sigma_\omega^2(t) = \omega_0(t)\sigma_0^2(t) + \omega_1(t)\sigma_1^2(t), \tag{6.7}$$

where ω_0 and ω_1 weights are the probabilities of the two classes separated by a threshold t, and σ_0^2 and σ_1^2 are the statistical variances of ω_0 and ω_1, respectively.

After the vessel segmentation procedure was applied, the Medial Axis Transform technique was used to extract the vessel skeletons, which makes use of the Voronoi method, expressed as follows:

$$R_k = \{x \in X | d(x, P_k) \le d(x, P_j) \text{ for all } j \neq k\}, \tag{6.8}$$

where R_k is the Voronoi region associated with the site P_k (a tuple of nonempty subsets in the space X), which contains the set with all points in X whose distance to P_k is not greater than their distance to the other sites P_j. j is any index different from k. $d(x, P_k)$ is a closeness measure from point x to point P_k. The Euclidean distance is commonly used as a closeness measure and it is defined as follows:

$$D(p_1, p_2) = \sqrt{(x_2 - x_1)^2 + (y_2 - y_1)^2} \tag{6.9}$$

where $D(p_1, p_2)$ is the distance between two points p_1 and p_2 defined by coordinates (x_1, y_1) and (x_2, y_2), respectively, in a 2-D plane.

Figure 6.6 illustrates five coronary angiograms with their respective segmentation response applying the Frangi method and the corresponding results after the binarization and skeletonization tasks were performed. In Section A.1 is presented a Matlab implementation to extract vessel segments. The code function makes use of other auxiliary functions presented in subsequent sections in the appendix. For instance, in Section A.2 is presented the code that finds and returns the positions related with the pixels of a segment in order to calculate their morphology on a posterior process. The code in Section A.3 is useful to extract a window from a 2-D matrix. Finally, the code presented in Section A.4 finds and returns the first row position of a matrix where a specific row vector is located. Those methods are useful to extract distinct features related with vessel's morphology.

6.2.2 FEATURE SELECTION

Feature selection refers to select only those features from a larger set that are useful to solve some specific problem. Since a significant amount of features can be

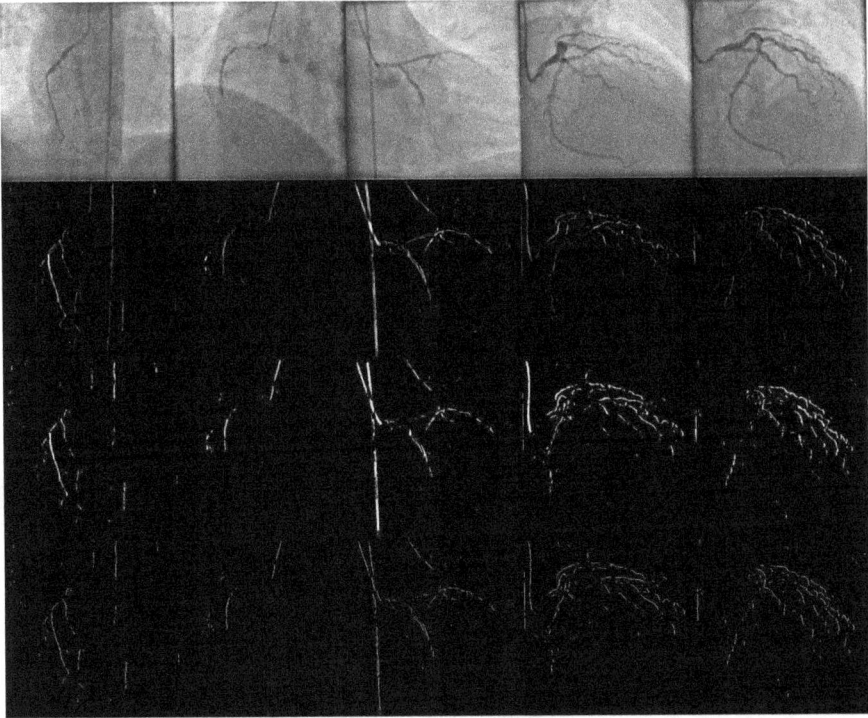

Figure 6.6: Image vessel detection. In first row, original angiograms are presented. Second row shows the Frangi segmentation response. In third row, the Otsu method response is illustrated. Last row contains the vessel skeletons by applying the Medial Axis Transform method.

extracted, the challenge is to find the best feature subset that is able to classify coronary stenosis and non-stenosis cases correctly. The complexity associated to solve the posed problem relies in the large number of combinations of features to achieve the best classification rate. The problem complexity can be computed as 2^n possibilities to combine features, where n is the number of features. Figure 6.7 illustrates two examples of feature selection.

6.2.3   SUPPORT VECTOR MACHINES

Support Vector Machines (SVMs) were designed primary to be linear classifiers by applying concepts of vector calculus in physics [16][17]. Major strength of SVMs relies in their capacity to project data to high dimensional orders than they are represented initially and, where a linear separation can be performed correctly [18]. In order to perform a linear classification, SVMs select a subset of data points that are able to influence the position and orientation of the hyperplane, which can be used to perform the linear classification. It is important to be mentioned that SVMs are

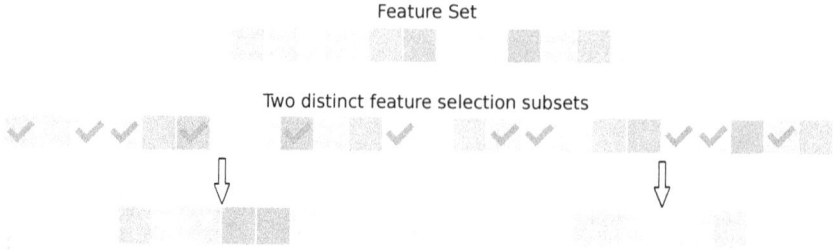

Figure 6.7: Feature selection example.

supervised learning models. This means that a training data set and its corresponding label set are required to perform the SMV training. In order to project the data represented in a space χ to a higher dimensional space $\mathscr{F}$, the SVM makes use of the Mercer kernel operator. For a given training data $x_1, ..., x_n$, that are vectors in some space $\chi \subseteq \mathbb{R}^d$, the support vectors can be considered as a set of classifiers expressed as follows [19]:

$$f(x) = \left(\sum_{i=1}^{n} \alpha_i K(x_i, x) \right). \tag{6.10}$$

When K satisfies the Mercer condition [20], it can be expressed as follows:

$$K(u, v) = \Phi(u) \cdot \Phi(v), \tag{6.11}$$

where $\Phi : \chi \to \mathscr{F}$ and "·" denotes an inner product. With this assumption, f can be rewritten as follows:

$$f(x) = w \cdot \Phi(x),$$
$$w = \sum_{i=1}^{n} \alpha_i \Phi(x_i). \tag{6.12}$$

6.2.4 QUANTUM GENETIC ALGORITHM

As described before in Section ??, feature selection is a complicated task when the number of features is considerably large, since the total number of different combinations of features can be computed as 2^n. For this reason, by performing random or manual trials will have low probabilities to find the optimal combination of features able to achieve an accurate classification rate and minimizing at same time, the number of involved features. In order to overcome this problem, the use of metaheuristic search strategies is a suitable way to explore. In this research, the Quantum Genetic Algorithm (QGA) was employed due to the advantages that it has over other search strategies.

QGA is the product of the combination of quantum computation and genetic algorithms, and it is a new evolutionary algorithm of probability [21]. In 1996, QGA was first proposed by Narayanan and Moore [22]. It was successfully used to solve the TSP problem [23]. QGA can be applied to solve problems where the conventional genetic algorithm has been applied. In the state of the art literature a wide

of QGAs implementations are described to solve optimization problems successfully [24][25][26][27][28], such as the personnel scheduling problem [29], dynamic
economic dispatch problem [30], multi-sensor image registration [31], cryptanalysis [32], and others.

Quantum computing is based on the *evolution* concept from quantum mechanics,
which establishes that the evolution of an isolated quantum system from an initial
state to another final state is governed by the Schrödinger equation [33], expressed
as follows:

$$i\hbar = \frac{\partial}{\partial t}|\psi(t)\rangle = H(t)|\psi(t)\rangle, \tag{6.13}$$

where i is the imaginary number $\sqrt{-1}$ and $\hbar$ is a constant term named "reduced
Planck constant" and $\hbar = h/2\pi$. The $|\psi(t)\rangle$ vector depends on time and describes the
state of a quantum system (or state) at time t. In addition, it can be noted that quantum
mechanics equations are expressed using proper notation symbols to describe vector
and matrix operations [34]. In 1930, the theoretical physicist Paul Dirac published the
book The Principles of Quantum Mechanics introducing the *ket* notation to describe
a column vector as follows:

$$|V\rangle = \begin{pmatrix} v_1 \\ v_2 \\ \vdots \\ v_i \end{pmatrix} \tag{6.14}$$

and *bra* notation to denote a row vector as following:

$$\langle M| = \begin{pmatrix} w_1 & w_2 & \cdots & w_i \end{pmatrix} \tag{6.15}$$

therefore, the product of bra and ket vectors can be expressed using the notation of
Dirac as follows:

$$\langle M|V\rangle = \begin{pmatrix} w_1 & w_2 & \cdots & w_i \end{pmatrix} \begin{pmatrix} v_1 \\ v_2 \\ \vdots \\ v_i \end{pmatrix} = w_1v_1 + w_2v_2 + \ldots + w_iv_i \tag{6.16}$$

The products of ket and bra vectors are expressed as follows:

$$|V\rangle\langle M| = \begin{pmatrix} v_1 \\ v_2 \\ \vdots \\ v_i \end{pmatrix} \begin{pmatrix} w_1 & w_2 & \cdots & w_i \end{pmatrix} = \begin{pmatrix} v_1w_1 & v_1w_2 & v_1w_{\cdots} & v_1w_i \\ v_2w_1 & v_2w_2 & v_2w_{\cdots} & v_2w_i \\ \vdots & \vdots & \vdots & \vdots \\ v_iw_1 & v_iw_2 & v_iw_{\cdots} & v_iw_i \end{pmatrix} \tag{6.17}$$

When translating previous physics concepts to computing field, the term *qubit* is used
to represent the minimal unit of information that stores $|0\rangle$ and $\langle 1|$ states. Applying
this principle to the QGA, an initial quantum population $Q(0)$ is composed of a set of
quantum individuals. Each individual will be composed of a quantum chromosome

i that is defined as a string of j qubits representing a quantum system $|\psi\rangle^i$ with 2^j simultaneous states, expressed as following:

$$\begin{pmatrix} \alpha_1 & \alpha_2 & \alpha_3 & \cdots & \alpha_j \\ \beta_1 & \beta_2 & \beta_3 & \cdots & \beta_j \end{pmatrix}_i \rightarrow |\psi\rangle^i = \sum_{k=1}^{j} c_i |\psi_k\rangle, \qquad (6.18)$$

being the gene j, the qubit represented by a vector expressed as follows:

$$\begin{pmatrix} \alpha_j \\ \beta_j \end{pmatrix} \rightarrow |\psi_j\rangle \qquad (6.19)$$

and the quantum population expressed as following:

$$\begin{pmatrix} \alpha_1 & \alpha_2 & \alpha_3 & \cdots & \alpha_j \\ \beta_1 & \beta_2 & \beta_3 & \cdots & \beta_j \end{pmatrix}_1$$
$$\begin{pmatrix} \alpha_1 & \alpha_2 & \alpha_3 & \cdots & \alpha_j \\ \beta_1 & \beta_2 & \beta_3 & \cdots & \beta_j \end{pmatrix}_2 \qquad (6.20)$$
$$\vdots$$
$$\begin{pmatrix} \alpha_1 & \alpha_2 & \alpha_3 & \cdots & \alpha_j \\ \beta_1 & \beta_2 & \beta_3 & \cdots & \beta_j \end{pmatrix}_i$$

In QGA the initial population is generated by setting the value for all qubits in the chromosomes by representing the quantum superposition of all states with equal probability [33]. The initialization can be computed using the Hadamard matrix by the vector $|0\rangle$ as follows:

$$H.|0\rangle = \frac{1}{\sqrt{2}} \begin{pmatrix} 1 & 1 \\ 1 & -1 \end{pmatrix} \begin{pmatrix} 1 \\ 0 \end{pmatrix} = \frac{1}{\sqrt{2}} \begin{pmatrix} 1 \\ 1 \end{pmatrix}, \qquad (6.21)$$

obtaining a superposition vector. After the product was computed, a phase angle θ $(0, \frac{\pi}{2}]$ is randomly obtained, being the argument of the trigonometric functions or elements in the rotation matrix $U(t)$, expressed as follows:

$$U(t) = \begin{pmatrix} \cos(\theta) & -\sin(\theta) \\ \sin(\theta) & \cos(\theta) \end{pmatrix} \qquad (6.22)$$

Finally, to conclude the initialization step, the product of the rotation matrix by the superposition vector must be performed, obtaining a pair of amplitudes (α, β), which is defined as the state of j qubit as follows:

$$\begin{pmatrix} \alpha_j \\ \beta_j \end{pmatrix} = \begin{pmatrix} \cos(\theta) & -\sin(\theta) \\ \sin(\theta) & \cos(\theta) \end{pmatrix} \frac{1}{\sqrt{2}} \begin{pmatrix} 1 \\ 1 \end{pmatrix} \qquad (6.23)$$

In order to obtain a discrete population of chromosomes, an *observation* must be performed as follows:

$$\begin{cases} p(\alpha) \leq |\alpha_j^2|, x_j = 0 (\text{basisstate} |0\rangle) \\ p(\alpha) > |\alpha_j^2|, x_j = 1 (\text{basisstate} |1\rangle) \end{cases} \qquad (6.24)$$

Such as in classic GA, on each iteration, crossover and mutation of population is performed in order to produce new individuals. These operations are known as Quantum Crossover Gate and Quantum Mutation Gate. The mutation occurs in two different manners: inversion and insertion. To perform the inversion-based mutation, an inter-qubit mutation of the jth qubit is computed by swapping the amplitudes with the quantum Pauli X gate as follows:

$$U(t) = \begin{pmatrix} 0 & 1 \\ 1 & 0 \end{pmatrix}, \tag{6.25}$$

which results in

$$\begin{pmatrix} \beta_j^{t+1} \\ \alpha_j^{t+1} \end{pmatrix} = \begin{pmatrix} 0 & 1 \\ 1 & 0 \end{pmatrix} \begin{pmatrix} \alpha_j^t \\ \beta_j^t \end{pmatrix}$$

The quantum mutation using the insertion gate is performed by the permutation or swapping between two qubits chosen randomly. For instance, given the following quantum chromosome with first and third qubits chosen randomly:

$$\begin{pmatrix} \alpha_1 & \alpha_2 & \alpha_3 & \cdots & \alpha_j \\ \beta_1 & \beta_2 & \beta_3 & \cdots & \beta_j \end{pmatrix},$$

the new mutated chromosome applying the insertion strategy will be as follows:

$$\begin{pmatrix} \alpha_3 & \alpha_2 & \alpha_1 & \cdots & \alpha_j \\ \beta_3 & \beta_2 & \beta_1 & \cdots & \beta_j \end{pmatrix}$$

The quantum crossover is performed similar to classic GA by selecting a randomly cut point and exchanging chromosomal segments. For instance, suppose that two chromosomes m and n expressed as follows:

$$\begin{pmatrix} \alpha_1 & \alpha_2 & \alpha_3 & \cdots & \alpha_j \\ \beta_1 & \beta_2 & \beta_3 & \cdots & \beta_j \end{pmatrix}_m$$

$$\begin{pmatrix} \alpha_1' & \alpha_2' & \alpha_3' & \cdots & \alpha_j' \\ \beta_1' & \beta_2' & \beta_3' & \cdots & \beta_j' \end{pmatrix}_n$$

were selected for a crossover operation with a randomly selected point between first and second positions. The resultant recombined chromosomes are expressed as follows:

$$\begin{pmatrix} \alpha_1 & \alpha_2' & \alpha_3' & \cdots & \alpha_j' \\ \beta_1 & \beta_2' & \beta_3' & \cdots & \beta_j' \end{pmatrix}_{m^*}$$

$$\begin{pmatrix} \alpha_1' & \alpha_2 & \alpha_3 & \cdots & \alpha_j \\ \beta_1' & \beta_2 & \beta_3' & \cdots & \beta_j \end{pmatrix}_{n^*}$$

Using the previous concepts, the QGA steps are defined and compared with its classical corresponding GA in Table 6.1.

Table 6.1

QGA Steps

Step	Quantum GA	Classic GA
1	Initialize quantum population Q_0	Generate initial population P_0
2	Make $P(0)$ by measuring each individual $Q(0) \rightarrow P(0)$	Evaluate $P(0)$
3	while (not termination condition) do	while (not termination condition) do
4	begin	begin
5	$t \leftarrow t + 1$	$t \leftarrow t + 1$
6	Perform Quantum Crossover	Perform Crossover
7	Perform Quantum Mutation	Perform Mutation
8	Measure $Q(t) \rightarrow P(t)$	Evaluate population $P(t)$
9	end	end

6.3   PROPOSED METHOD

This research is focused in the use of a QGA to select a subset of features that are suitable to perform the classification of stenosis and non-stenosis cases keeping or improving at same time, the accuracy rate that is obtained by using the original feature set. The method starts with a feature extraction task in order to obtain an initial set of features from a bank of image patches. In this first step, distinct types of features are extracted such as basic-intensity measures, texture-based and shape features. By performing this step, an initial set of 31 features are extracted from the image bank. Features 1 to 14 correspond to Haralick texture features described as follows:

1. Angular second moment (energy)
2. Contrast
3. Correlation
4. Variance
5. Inverse difference moment (homogeneity)
6. Sum average
7. Sum variance
8. Sum entropy
9. Entropy
10. Difference variance
11. Difference entropy
12. Information measure of correlation 1
13. Information measure of correlation 2
14. Maximum correlation coefficient

Features 15 to 18 correspond to basic intensity measurements in the image.

15. Minimum intensity value
16. Maximum intensity value
17. Mean intensity of the image
18. Standard deviation of the intensities in the image

The remaining features are related with the vessel morphology. As mentioned before, a previous vessel enhancement and skeletonization tasks are required to keep only information related with vessel pixels. The Frangi method was applied in order to perform a vessel enhancement over the original images. Later, a binarization of each Frangi response was performed by applying the Otsu method in order to discriminate non-vessel pixels. Finally, the Medial Axis Transform procedure was applied in order to extract vessel skeletons. Shape-based features are described as follows:

19. The number of pixels corresponding to vessels in the patch
20. The number vessel segments in the patch
21. Vessel density. The proportion of vessel-pixels present in the patch
22. Tortuosity 1. The tortuosity of each segment was calculated using the true length (measured with the chain code) divided by the Euclidean length. The mean tortuosity was calculated from all the segments within the sub window
23. Vessel Length
24. Number of bifurcation points. The number of bifurcation points removed within the sub window when creating segments.
25. Gray level coefficient of variation. The ratio of the standard deviation to the mean of the grey level of all segment pixels within the sub window.
26. Gradient mean. The mean gradient magnitude along all segment pixels within the sub window calculated using the Sobel gradient operator applied on the pre-processed image.
27. Gradient coefficient of variation. The ratio of the standard deviation to the mean of the gradient of all segment pixels within the sub window.
28. Mean vessel width. Skeletonization correlates to vessel center lines. The distance from the segment pixel to the closest boundary point of the vessel using the vessel map prior to skeletonization. This gives the half-width at that point, which is then multiplied by 2 to achieve the full vessel width. The mean is calculated for all segment pixels within the sub-window
29. The minimum standard deviation of the vessels length, based on the vessels present in the patch
30. The maximum standard deviation of the vessels length, based on the vessels present in the patch
31. The mean of the standard deviations of the vessels length, based on the vessels present in the patch

It is important to mention that a vessel segment is referred to the skeletonized vessel segments. After a feature set is defined, a QGA is used with an SVM in order to search for an optimal feature selection. The proposed method procedure is described in detail in Figure 6.8.

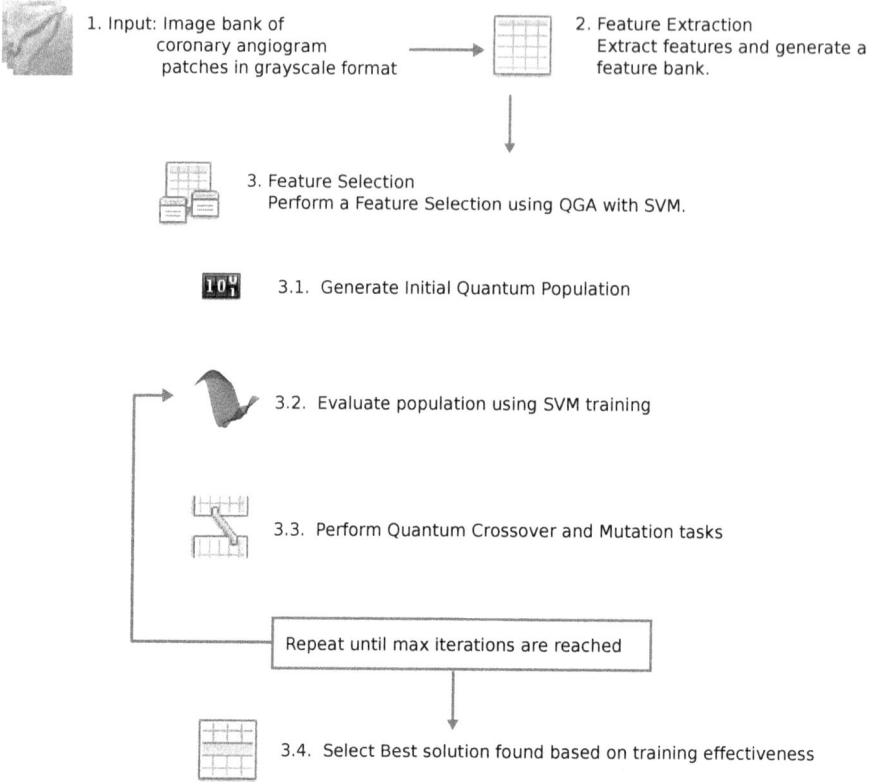

1. Input: Image bank of
   coronary angiogram
   patches in grayscale format

2. Feature Extraction
   Extract features and generate a
   feature bank.

3. Feature Selection
   Perform a Feature Selection using QGA with SVM.

3.1. Generate Initial Quantum Population

3.2. Evaluate population using SVM training

3.3. Perform Quantum Crossover and Mutation tasks

Repeat until max iterations are reached

3.4. Select Best solution found based on training effectiveness

Figure 6.8: Feature selection example.

After the feature selection task is finished, the SVM can be trained using only the feature subset found previously by the QGA. In order to probe the effectiveness, a testing set is used. In addition, the classification results are measured using two different metrics in order to assess the achieved results. As a first instance, the True-Positive (TP), True-Negative (TN), False-Negative (FN), and False-Positive (FP) fractions are used to obtain the Accuracy (Acc) metric. This metrics are computed as follows:

$$TPR = \frac{TP}{TP+FN} \quad , \tag{6.26}$$

$$TNR = \frac{TN}{TN+FP} \quad , \tag{6.27}$$

$$FPR = \frac{FP}{FP+TN} \quad , \tag{6.28}$$

$$Acc = \frac{TP+TN}{TP+TN+FP+FN} \quad , \tag{6.29}$$

where TPR is the True-Positive Rate, TNR is the True-Negative Rate, FPR is the False-Positive Rate, TP is the number of positive instances that were labeled as positive by the classifier, TN is the number of negative instances labeled as negative by the classifier, FP is the number of negative instances labeled as positive by the classifier, and FN is the number of positive instances labeled as negative by the classifier.

As a second accuracy metric, the Jaccard index is used. The Jaccard index is a measure that determines how similar are two sets of elements. It is useful to measure the efficiency of a classifier by comparing the obtained results versus those expected. It is computed as follows:

$$J(A,B) = \frac{A \cap B}{A \cup B},$$ (6.30)

where A and B are the expected and response sets, respectively.

6.4 EXPERIMENT DETAILS

For the experimentation, an image bank of coronary patches was formed. The images correspond to the natural Antczak [8] dataset, which is formed by 1394 negative and 122 positive instances. However, in order to avoid an overfitting of the SVM, 1272 positive instances were taken from the Antczak synthetic dataset. With the synthetic positive-stenosis patches addendum, the dataset was equilibrated with 1394 positive cases and 1394 negative cases. Each patch is of size 32×32 pixels and they are defined in gray-scale. Figure 6.9 illustrates a set of samples with positive and negative cases.

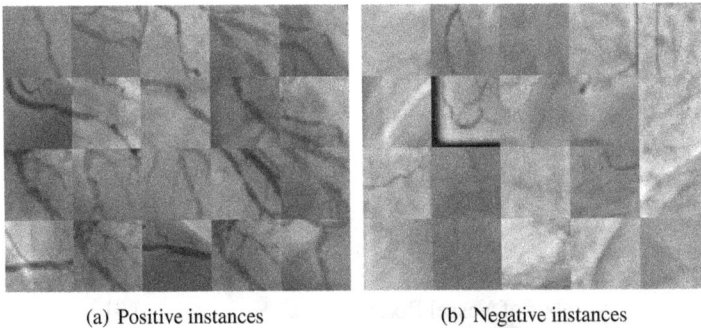

(a) Positive instances   (b) Negative instances

Figure 6.9: 40 instances of coronary X-ray angiogram patches taken from the Antczak [8] image dataset. (a) Positive stenosis cases. (b) Negative stenosis cases.

For the training and the feature selection process, 1670 instances were taken randomly from the image set in a proportion of $50 - 50\%$ for positive and negative cases, respectively. The remaining instances were used for testing purposes.

As mentioned before, an SVM was used to perform the classification of positive and negative cases. It was configured to use a polynomial kernel of order 6, a kernel scale of 6.5, and a kernel offset of 0.1. All SVM parameters were established manually as a trade-off between time and accuracy. Additionally, in order to increase the train rate of the SVM, the training dataset was partitioned into 10 groups with 167 instances in order to perform a cross-validation with $k = 10$.

All experiments were performed using the Matlab 2018Rb platform using an Intel i3 PC with 4GB of RAM.

6.5 RESULTS

After conducting the experiment, significant results were obtained related with the feature selection process and the classification accuracy achieved. In order to assess the feature selection results, a statistical analysis was performed considering the best result achieved by the QGA on each trial from a set of 30 trials. In its best global solution, the QGA was able to find an optimal feature subset of 20 instead of the initial set containing 31 features. Additionally, it is interesting to notice the frequency which each feature was present in the best solution achieved by QGA over all trials, since it provides information about the influence of that particular feature on the classification result. Table 6.2 describes the frequency which each feature was present in the best solution achieved by the QGA in each trial.

Table 6.2

Feature Selection Frequency over All QGA Trials

Feature	Frequency
Contrast	0.55
Correlation	0.64
Sum Variance	0.73
Sum Entropy	0.45
Entropy	0.45
Difference Entropy	0.45
Min	1.00
Max	1.00
Mean	0.64
Std.Dev.	0.91
Number of Vessel Pixels	0.91
Number of Vessel Segments	0.73
Vessel Density	0.91
Vessel Length	0.91
Number of Bifurcation Points	0.64
Grey Level Coefficient of Variation	1.00
Gradient Mean	0.73
Gradient Coefficient of Variation	0.91
Mean Vessel Width	0.91
Vessel. Std. Dev. Max	0.64

Figure 6.10 illustrates the frequency for all set of 31 features. Based on the results

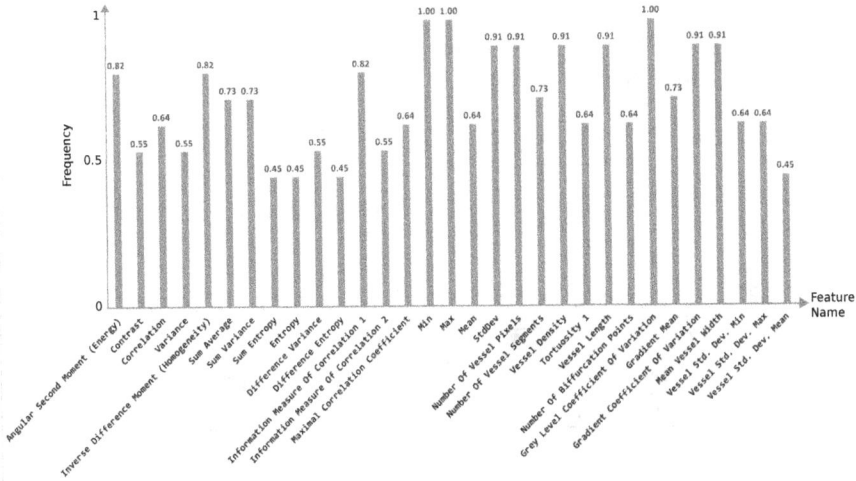

Figure 6.10: Frequency in which each feature was present in best solution achieved by QGA over all trials. *x*-axis describes each feature name. *y*-axis represents the frequency of each feature, where 1 means that the feature appears in the solution achieved by the QGA in all trials (the 100%).

described in Table 6.2 and the chart presented in Figure 6.10, the importance or effect of each individual feature in the classification task and how the combination of them can lead to an optimal classification rate can be observed. It is remarkable that some features appear on all solutions achieved by the QGA, meaning that those features are really important in the classification process. In addition, it can be contrasted the results described in Table 6.2 with Figure 6.10 and how almost a half of selected features are above of the mean frequency, which was 0.72.

The performance results are presented in Table 6.3. The proposed method is compared with other four classification methods in the literature, such as Feed Forward (FFNN) and Back Propagation (BPNN) Neural Networks, UNET and CNN-16C Convolutional Neural Networks.

Based on results presented in Table 6.3, the classification process was performed with the highest rates in terms of the Accuracy and Jaccard Index metrics by the SVM-based classification method. However, the proposed method was able to achieve closest performance results to the SVM by using only a subset of 20 features instead of the total feature set, which contains a total of 31 features. It is important to notice that the performance of non-linear classification methods such as FFNN and BPNN decreases when a feature selection task is applied to them. This can be attributed to the loss of information associated with the elimination of features which are relevant for a non-linear classification task. On the other hand, deep learning methods such as UNET and CNN-16C Convolutional Neural Networks were able to achieve the closest results to the best rate after the proposed method. In this

Table 6.3

Accuracy Rate and Jaccard Index Comparison for the Proposed Method and 5 additional Classification Methods from the State of the Art

Method	Number of Features	Accuracy	Jaccard Index
FFNN	20	0.70	0.54
	31	0.72	0.61
BPNN	20	0.69	0.57
	31	0.71	0.00
SVM	31	**0.92**	**0.86**
UNET [7]	–	0.76	0.72
CNN-16C [8]	–	0.86	0.74
Proposed Method	**20**	**0.92**	0.85

context, it is important to mention that CNNs were applied using only the same training and testing set such as the rest of techniques and no data-augmentation or transfer-learning processes were applied in order to measure the effectiveness of all techniques under the same conditions. In Figure 6.11, a subset of instances corresponding to the *True-Positive*, *True-Negative*, *False-Positive*, and *False-Negative* fractions are illustrated.

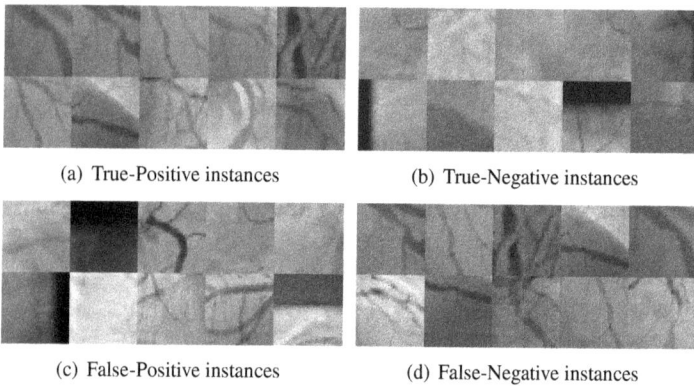

(a) True-Positive instances

(b) True-Negative instances

(c) False-Positive instances

(d) False-Negative instances

Figure 6.11: Samples of the true-positive, true-negative, false-positive, and false-negative, fractions of the accuracy metric. Each fraction contains 10 instances.

6.6   CONCLUSION

In this research, a method is proposed for the stenosis classification problem, which makes use of a Quantum Genetic Algorithm (QGA) to perform an automatic feature selection in order to select only those features that has a strong influence on a Support Vector Machine-based (SVM) classifier. The QGA performs a search over the space formed by the feature set in order to find an optimal combination of features and at the same time, keeping or decreasing the loss rate in the training stage. Initially, a set of 31 features were extracted from an image database which contains ≈ 2780 instances of X-ray coronary angiogram patches. The image database is balanced in terms of the positive and negative stenosis cases. After the feature selection process ends, a subset of 20 features was able to keep the classification rate in terms of the Accuracy metric and the Jaccard index, compared with the original set with 31 features. By using only 20 features, the accuracy rate and Jaccard index were 0.92 and 0.85, respectively, which are very similar to those obtained using the full set of 31 features. In addition, the reduction of features has effect on the time required to perform an exhaustive feature extraction of new angiograms, since the required time for extract the 31 features was 0.94 seconds, versus 0.62 seconds that are required to extract the 20 selected features, considering a window that will perform a scan over an angiogram to detect regions with possible stenosis cases. Based on the results obtained in this study, it can be concluded that the proposed method can be applied in clinical practice to assists cardiologists in the evaluation and finding of possible stenosis cases in X-ray coronary angiograms.

ACKNOWLEDGMENT

The present research has been supported by the Universidad Tecnológica de León.

REFERENCES

1. A. Kishore and V. Jayanthi. Automatic stenosis grading system for diagnosing coronary artery disease using coronary angiogram. International Journal of Biomedical Engineering and Technology, 31(3):260–277, 2018.
2. I. Saad. Segmentation of coronary artery images and detection of atherosclerosis. Journal of Engineering and Applied Sciences, 13:7381–7387, 2018.
3. S. Sameh, M. Azim, and A. AbdelRaouf. Narrowed coronary artery detection and classification using angiographic scans. In 2017 12th International Conference on Computer Engineering and Systems (ICCES), pages 73–79, 2017.
4. T. Wan, H. Feng, C. Tong, D. Li, and Z. Qin. Automated identification and grading of coronary artery stenoses with x-ray angiography. Computer Methods and Programs in Biomedicine, 167:13–22, 2018.
5. F. Cervantes-Sanchez, I. Cruz-Aceves, and A. Hernandez-Aguirre. Automatic detection of coronary artery stenosis in x-ray angiograms using Gaussian filters and genetic algorithms. AIP Conference Proceedings, 1747, 2016.
6. I. Cruz-Aceves, F. Cervantes-Sanchez, and A. Hernandez-Aguirre. Automatic detection of coronary artery stenosis using Bayesian classification and Gaussian filters based on differential evolution. Hybrid Intelligence for Image Analysis and Understanding, pages 369–390, 2017.
7. A. Harouni, A. Karargyris, M. Negahdar, D. Beymer, and T. Syeda-Mahmood. Universal multi-modal deep network for classification and segmentation of medical images. In 2018

IEEE 15th International Symposium on Biomedical Imaging (ISBI 2018), pages 872–876, 2018.

8. Karol Antczak and ukasz Liberadzki. Stenosis detection with deep convolutional neural networks. MATEC Web of Conferences, 210:04001, 2018.

9. Shin Hoo-Chang, Holger Roth, Mingchen Gao, Le Lu, Ziyue Xu, Isabella Nogues, Jianhua Yao, Daniel Mollura, and Ronald-M. Summers. Deep convolutional neural networks for computer-aided detection: CNN architectures, dataset characteristics and transfer learning. IEEE Transactions on Medical Imaging, 35(5):1285–1298, 2016.

10. Hossein Azizpour, Ali Razavian, Josephine Sullivan, Atsuto Maki, and Stefan Carlsson. From generic to specific deep representations for visual recognition. In 2015 IEEE Conference on Computer Vision and Pattern Recognition Workshops (CVPRW), pp. 36–45, 2015.

11. David-A. van Dyk and Xiao-Li Meng. The art of data augmentation. Journal of Computational and Graphical Statistics, 10(1):1–50, 2001.

12. Jessamyn Dahmen and Diane Cook. Synsys: A synthetic data generation system for healthcare applications. Sensors, 19(5): 2019.

13. R. Haralick, K. Shanmugam, and I. Dinstein. Textural features for image classification. IEEE Transactions on Systems, Man, and Cybernetics, 3(6):610–621, 1973.

14. A. Frangi, W. Niessen, K. Vincken, and M. Viergever. Multiscale vessel enhancement filtering. Medical Image Computing and Computer-Assisted Intervention (MICCAI98), pages 130–137, 1998.

15. Otsu Nobuyuki. A threshold selection method from gray-level histograms. IEEE Transactions on Systems, Man and Cybernetics, 9(1):62–66, 1979.

16. Corinna Cortes and Vladimir Vapnik. Support-vector networks. Machine Learning, 20:273–297, 1995.

17. Nello Cristianini, John Shawe-Taylor, et al. An introduction to support vector machines and other kernel-based learning methods. Cambridge University Press, 2000.

18. William-S. Noble. What is a support vector machine? Nature Biotechnology, 24:1565–1567, 2006.

19. Simon Tong and Daphne Koller. Support vector machine active learning with applications to text classification. Journal of Machine Learning Research, pages 45–66, 2001.

20. Christopher-J.C. Burges. A tutorial on support vector machines for pattern recognition. Data Mining and Knowledge Discovery, 2:121–167, 1998.

21. F. Shi, H. Wang, L. Yu, and F. Hu. Analyze of 30 Cases of MATLAB Intelligent Algorithms. Beihang University Press, 2010.

22. A. Narayanan and M. Moore. Quantum-inspired genetic algorithms. In Proceedings of IEEE International Conference on Evolutionary Computation, pages 61–66, 1996.

23. Davendra Donald. Traveling salesman problem, theory and applications. Intech, Rijeka, 2011.

24. Kuk-Hyun Han and Jong-Hwan Kim. Quantum-inspired evolutionary algorithms with a new termination criterion, h/sub /spl epsi// gate, and two-phase scheme. IEEE Transactions on Evolutionary Computation, 8(2):156–169, 2004.

25. Zhifeng Zhang and Hongjian Qu. A new real-coded quantum evolutionary algorithm. In Proceedings of the 8th WSEAS International Conference on Applied Computer and Applied Computational Science, pages 426–429, 2009.

26. Gexiang Zhang. Quantum-inspired evolutionary algorithms: A survey and empirical study. Journal of Heuristics, 17:303–351, 2011.

27. Utpal Roy, Sudarshan Roy, and Susmita Nayek. Optimization with quantum genetic algorithm. International Journal of Computer Applications, 102(16):1–7, 2014.

28. Ying Sun and Xiong Hegen. Function optimization based on quantum genetic algorithm. Research Journal of Applied Sciences, Engineering and Technology, 7(1):144–149, 2014.

29. Wang Huaixiao, Li Ling, Liu Jianyong, Wang Yong, and Fu Chengqun. Improved quantum genetic algorithm in application of scheduling engineering personnel. Abstract and Applied Analysis, 2014:1–10, 2014.

30. Lee Jia-Chu, Lin Whei-Min, Liao Gwo-Ching, and Tsao Ta-Peng. Quantum genetic algorithm for dynamic economic dispatch with valve-point effects and including wind power system. International Journal of Electrical Power & Energy Systems, 33(2):189–197, 2011.

31. Ying Sun and Xiong Hegen. A novel quantum-inspired evolutionary algorithm for multisensor image registration. The International Arab Journal of Information Technology, 3(1):9–15, 2006.

32. Hu Wei. Cryptanalysis of tea using quantum-inspired genetic algorithms. Journal of Software Engineering and Applications, 3:50–57, 2010.

33. Rafael Lahoz-Beltra. Quantum genetic algorithms for computer scientists. Computers, 5(4):243–249, 2012.

34. Leonard Susskind and Art Firedman. Quantum Mechanics: The Theoretical Minimum. Penguin Books: London, UK, 2015.

7 Quantum Preprocessing for Deep Convolutional Neural Networks in Atherosclerosis Detection

7.1 INTRODUCTION

Atherosclerosis is a specific type of stenosis, i.e., narrowing or occlusion of the artery lumen, that occurs due to the accumulation of some substances as cholesterol on the coronary arteries' inner walls. An opportune diagnosis and atherosclerosis treatments are essential since it represents the leading cause of Coronary Artery Disease. According to the World Health Organization (WHO), this heart condition has a high mortality rate worldwide, with 17.9 million estimated deaths every year [1]. Atherosclerosis detection consists of arteries visual inspection through a screening test, either non-invasively employing computed tomography (Coronary Computed Tomography Angiography, CCTA) or with the regular procedure consisting of inserting a catheter through the groin or arm into the coronary arteries (X-ray Coronary Angiography, XCA). In both cases, a contrast medium is injected to guide or locate the arteries. Nonetheless, XCA remains the gold standard used by specialists since it offers enough resolution for diagnosis. Besides, the patients can receive treatment during the same session. For instance, during the visual examination of the XCA images, the physician may identify regions with stenosis as shown in Figure 7.1.

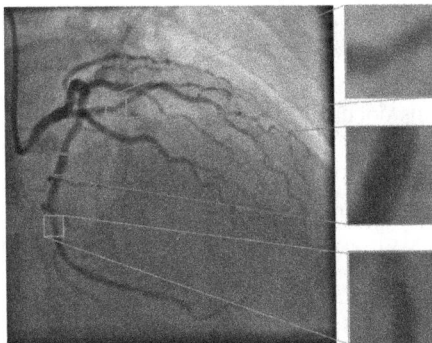

Figure 7.1: X-ray coronary angiography image. Stenosis and non-stenosis samples regions marked in green and red, respectively.

DOI: 10.1201/9781003283294-7

However, given the limited access of specialists and the time consumption for diagnoses, it has allowed the Computer-Aided Diagnosis (CAD) systems to play a vital role in cardiology. CAD systems have been a significant field of research during the last few decades, developed to improve and support the medical diagnosis process. CAD uses Machine Learning (ML) methods to analyze imaging or non-imaging (i.e., clinical profile) patient data to assess patients' conditions.

In computer vision, one of the breakthroughs occurred when Krizhevsky *et al.* [2] presented *Alexnet*, a Deep Convolutional Neural Network (DCNN) that won the 2012 ImageNet Large Scale Visual Recognition Challenge (ILSVRC) [3]. New Deep Learning (DL) algorithms have been proposed to adapt CNN architectures to challenging medical imaging problems. Despite significant research problems on the medical imaging domain using DL algorithms, it often suffers two significant difficulties in practice: 1) limited amount of labeled data; 2) mislabeling labels. New hybrid models have emerged to build more efficient ML algorithms. The most relevant is Quantum Machine Learning (QML) [4], which relies on quantum computing emulated from physical properties. Such improvement can be accomplished using quantum mechanical systems.

In this work, a Hybrid Quantum-Convolutional Neural Network for atherosclerosis detection in XCA images is presented. This approach considers an automatic preprocessing step, implemented through a Quantum Convolutional Layer (QCL), whose behavior corresponds to apply a quantum circuits as an image filter. The prepossessing quantum layer receives as input an XCA patch of one single-channel. Moreover, it generates a multichannel image, which feeds a classical CNN to perform the atherosclerosis detection. The presented hybrid method showed that employing a QCL as an XCA image preprocessing improves the network performance against raw-XCA images, usually feeding a CNN. Five different evaluation metrics were used to measure the performance of the proposed method. Besides, two different optimization techniques were compared: Stochastic Gradient Descent and Stochastic Gradient Descent with Momentum. Additionally, two different CNN architectures previously introduced for atherosclerosis detection were studied. The employed dataset consists of 250 real XCA images, where 125 images are used for training and the remaining 125 for testing.

The remainder of this document is organized as follows: Firstly, in Section 7.2, the related work is addressed. Section 7.3 describes briefly the concepts related to Quantum computation and CNNs. Section 7.4 presents the introduced methodology for atherosclerosis detection. In Section 7.5, the experimental and numerical results are carefully detailed and discussed. Finally, the conclusions are stated in Section 7.6.

7.2 RELATED WORK

Some methods for the automatic detection of atherosclerosis have been proposed in the literature. Most of them have based their strategies on blood vessel extraction. Sui *et al.* [5] proposed performing vessel segmentation through a method including a multi-scale Hessian matrix-based enhancement and a dual-stage region growing method. The segmentation was used as input of an arterial tree's skeleton algorithm to compute the arterial diameter of each skeleton point. Finally, the detection of general or severe obstruction was assigned based on the ratio given by the segment minimum diameter over its average diameter for each segment.

Wu *et al.* [6] used a U-Net architecture to segment the vascular structure of the XCA sequence taking advantage of the binary output to calculate a contrast-filling degree for each frame. Next, a percentage of sequences passing to the detection step was selected, where a Deconvolutional Single-Shot Multibox Detector may generate the location information. Jevitha *et al.* [7] combined a contour activation algorithm with a Frangi filter to extract the left main coronary artery. Next, bifurcation points were automatically detected using a template kernel. Later, measurements were taken on the artery bifurcation angles. The result was used to discriminate between *stenotic* and healthy arteries. These methods building the atherosclerosis detection and classification ruled around blood vessel segmentation try to emulate physicians' procedures when diagnosing patients. This last approach has the advantage that the workflow often has a direct interpretation. Regardless, the main task results (atherosclerosis detection) are conditioned by the segmentation algorithm performance, which, in many cases, struggles to detect narrowed regions.

Recently, proposals have emerged that do not consider vessels' extraction as a necessary step. Instead, end-to-end systems have been built, taking advantage of the rapid increase in computational power and the superior performance that algorithms based on deep learning have shown.

Au *et al.* [8] introduced a patch-based CNN that automatically characterizes and analyzes coronary stenosis in such a context. The network was based on a DenseNet [9], where skip connections between convolutional layers were added. Likewise, Antczak and Liberadzki [10] developed a patch-based CNN based on a VGG architecture [11], adding dropout layers after the convolution's layers and removing the pooling layers. This approach employed an artificial dataset to overcome the problem of a limited amount of training data. Both networks were trained from scratch using a Sigmoid activation function in the last dense layer to compute the patch probability of belonging to the stenosis class.

Cong *et al.* [12] presented another interesting solution. The InceptionV3 [13] network was trained from scratch to select a subset of candidate frames (like what was done in [6]), considering the contrast filling degree and other image quality measures, such as well-defined vessel borders. A second step used Transfer Learning for stenosis detection with an InceptionV3 network pre-trained with the ImageNet database. Fine-tuning was performed using a strategy called redundancy training that included pre-classified redundant frames in the training dataset.

Ovalle *et al.* [14] presented a method for successfully detecting coronary artery stenosis in XCA images, evaluating three pre-trained state-of-the-art architectures (VGG16 [11], ResNet50 [15], and InceptionV3 [13]) via Transfer Learning. Such a method incorporates a network-cut approach where only a sub-set of layers was considered. Layers between the cut layer and custom classifier are discarded. During the fine-tuning step, an artificial dataset was exploited. Moreover, the fine-tuning was substantially improved using a sub-set of real XCA images. This approach outperformed the results obtained using only the fine-tuning process during Transfer Learning.

Figure 7.2 shows an overview of the previously published methods evaluated using XCA images. End-to-end methods have shown worthy potential for detecting

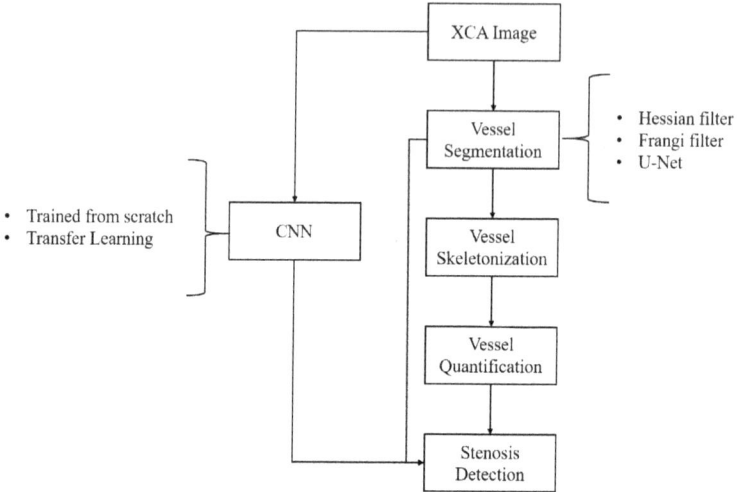

Figure 7.2: Typical workflow of the previously published algorithms for coronary artery stenosis detection in XCA images.

and classifying atherosclerosis using independent vascular structure extraction techniques. Besides, innovative visualization techniques have been efficiently proposed to interpret and justify the decision rules.

However, one of the most significant limitations of the previous approach is that a long training time is required to get accurate results. The number of model parameters stands in the order of thousands to millions. New hybrid models have then emerged to build more efficient machine learning algorithms. The Quantum Convolutional Layer or Quanvolutional Layer is an outstanding example of this approach [16]. It improves the performance of the CNN through quantum circuits representing classical filters. Quantum circuits can transform selected local subsections of the input data to extract relevant features, like traditional convolutional transformations whose filters are convolved into every local subsection of the input. Sleeman *et al.* [17] introduced a hybrid method combining a classical DCNN autoencoder with a quantum annealing Restricted Boltzmann Machine (RBM), applied to image generation. This hybrid autoencoder approach appeared an advantage for RBM results relative to using a classical RBM implementation into the MNIST (Modified National Institute of Standards and Technology) and MNIST Fashion dataset classification problem.

Furthermore, some relevant classification works related to the biomedical images domain have emerged. Iyer *et al.* [18] used a modified version of the Variational

Classifier previously introduced by Bergholm *et al.* in [19] for the binary classification of pigmented skin lesions. Images' dimensionality is transformed from $128 \times 128 \times 3$ to a 4-dimensional vector via an autoencoder cleverly being implemented with a CNN. Next, each vector was encoded in quantum amplitudes and sent to the 2-qubit classifier to obtain the predicted expected values of the image, being either a melanoma or melanocytic nevi.

On the other hand, Bisarya *et al.* [20] proposed a quantum convolutional neural network (QCNN) for cancer detection in breast cell data by exploiting the Wisconsin dataset. Two cases were theoretically considered: the former uses numerical data associated with the size of the cells and their texture, which were processed using an architecture of two QCLs with only ten parameters, each one assigned to a qubit; the latter uses gray-scale magnetic resonance images, processed in sets of 4×4-pixel arrays, encoded in a 256-qubits system. Regardless, only the first approach was implemented due to the limited availability of resources.

It is noteworthy that according to the authors' best knowledge, there are currently no works that use quantum convolutional networks to detect atherosclerosis or other types of stenosis. Therefore, in this study, a Hybrid Quantum-Convolutional Neural Network for atherosclerosis detection in XCA images, employing a QCL, is presented. The QCL successfully performs a preprocessing over the XCA images to generate a multichannel image feeding a classical CNN and improving the detection performance against the raw-XCA images.

7.3 MATHEMATICAL FOUNDATIONS

A summary of essential and associated topics of Quantum Computation and CNNs related to the proposed study is presented in the following subsections. The reader may refer to Nielsen and Chuang [21] and De Wolf [22] for in-depth details about quantum computation.

7.3.1 QUANTUM COMPUTING

The most straightforward computational unit is the *bit*, which can hold two different states, denoted as 0 and 1. On the other hand, the *qubit* is the basic unit in Quantum Computation [21]. Like a classical bit, a qubit also has two states, the *ground state* $|0\rangle$ and the *excited state* $|1\rangle$ (both written in Dirac notation). In contrast to classical bits whose value is a scalar, that is, zero or one, a qubit can be an arbitrary superposition of two basic states $|0\rangle$ and $|1\rangle$. Such a superposition might be interpreted as a linear combination of two basic vectors, weighted by complex amplitudes. This property is commonly known as *superposition*. Hence, these states can be manipulated through quantum circuits. Finally, a well-defined value can be obtained by the measurement of some (or all) of the qubits' output by the quantum circuit.

7.3.1.1 Qubit States

The qubit states are stochastic in nature due to the possibility of holding the superposition states. The two basic states are given as

$$|0\rangle = \begin{bmatrix} 1 \\ 0 \end{bmatrix}, \quad |1\rangle = \begin{bmatrix} 0 \\ 1 \end{bmatrix}. \tag{7.1}$$

The superposition state $|\psi\rangle$ is a linear combination of the basic states denoted by

$$|\psi\rangle = \alpha_0 |0\rangle + \alpha_1 |1\rangle, \tag{7.2}$$

where α_0 and α_1 are two complex numbers fulfilling the fundamental property given by $|\alpha_0|^2 + |\alpha_1|^2 = 1$. Thus, each amplitude α_i can be represented with the unit imaginary circle using the polar representation for a complex number. It can be shown that $|\psi\rangle$ can be written as

$$|\psi\rangle = \cos\left(\frac{\vartheta}{2}\right)|0\rangle + e^{i\varphi}\sin\left(\frac{\vartheta}{2}\right)|1\rangle, \tag{7.3}$$

where ϑ and φ correspond to two angles in spherical coordinates. In such a way, a quantum state is visualized as a unit vector in the 3-D space (on the Bloch sphere). The two basic states $|0\rangle$ and $|1\rangle$ are then represented by $\vartheta = 0$ and $\vartheta = \phi$, pointing in z and $-z$ directions, respectively, as illustrated in Figure 7.3.

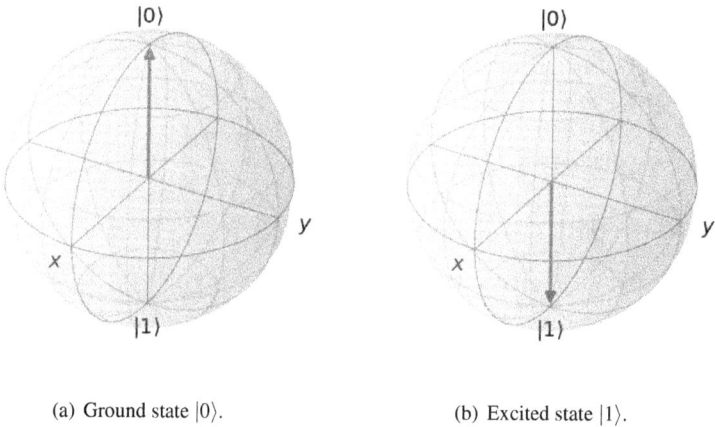

(a) Ground state $|0\rangle$.   (b) Excited state $|1\rangle$.

Figure 7.3: Illustration of the two basic states on the Bloch Sphere. The graph was plotted using the Johansson *et al.* library [23].

Next, a couple of examples are included to clarify some basic properties of quantum computing.

Example 1. *Consider a 2-qubit system whose state is described by the state vector* $\frac{1}{\sqrt{2}}[1,0,0,1]^\mathsf{T} = \frac{1}{\sqrt{2}}|0\rangle + \frac{1}{\sqrt{2}}|3\rangle = \frac{1}{\sqrt{2}}|(00)_2\rangle + \frac{1}{\sqrt{2}}|(11)_2\rangle$. *This representation is a* *valid quantum state since $|\frac{1}{\sqrt{2}}|^2 + |\frac{1}{\sqrt{2}}|^2 = 1$, where $(X)_2$ is a binary representation reading from left to right of the vector state positions.*

A quantum register or a quantum system of n-qubits can be any linear combination such that

$$|\psi\rangle_n = \alpha_0|0\rangle + \alpha_1|1\rangle + \cdots + \alpha_{2^n-1}|2^n-1\rangle,$$

$$\sum_{i=0}^{2^n-1}|\alpha_i|^2 = 1. \tag{7.4}$$

The *entanglement* is another essential property (unique to quantum computing). It means that a system of n-qubits cannot be written as the tensor product of two subsystems, e.g., a bipartite pure state containing the subsystems A and B, is entangled if it cannot be expressed as

$$|\psi\rangle_{AB} = |\varphi\rangle_A \otimes |\phi\rangle_B, \tag{7.5}$$

where $(\otimes)$ is the tensor product operator.

Example 2. *For a 2-qubit system, whose state is given by*

$$\frac{1}{\sqrt{2}}(|(00)_2\rangle + |(01)_2\rangle) = |0\rangle \otimes \frac{1}{\sqrt{2}}(|0\rangle + |1\rangle)$$

is separable since it can be represented as the tensor product of two subsystems $|\varphi\rangle_A = |0\rangle$ and $|\phi\rangle_B = \frac{1}{\sqrt{2}}(|0\rangle + |1\rangle)$. On the other hand, the system

$$\frac{1}{\sqrt{2}}(|(00)_2\rangle + |(11)_2\rangle) \neq |\varphi\rangle_A \otimes |\phi\rangle_B,$$

is entangled. Such a 2-qubit system is also known as an EPR-pair [24].

7.3.1.2 Qubit Operations

Quantum operations (or *quantum gates*) can manipulate the state of a quantum system such that

$$|\psi'\rangle = U|\psi\rangle, \tag{7.6}$$

where $U \in \mathbb{C}^{N \times N}$. Hence, it must be a unitary transformation given as

$$U^*U = UU^* = I, \tag{7.7}$$

where U^* is the conjugate transpose of U. A unitary operator acting on a small number of qubits (say, at most 3) is often called a gate, in analogy to the classical logic gates like AND, OR, and NOT. The simplest single-qubit gates consist of only two well-known gates: the Hadamard H gate and the T gate (or $\frac{\pi}{8}$ gate). The Hadamard gate allows taking a qubit from a definite computational basis state into a two states' superposition. The Hadamard matrix $\mathbf{H}$ is defined by

$$\mathbf{H} = \frac{1}{\sqrt{2}} \begin{bmatrix} 1 & 1 \\ 1 & -1 \end{bmatrix}. \tag{7.8}$$

Meanwhile, the $\mathbf{T}$-gate performs a rotation $\frac{\pi}{4}$ around the z-axis and is defined as follows

$$\mathbf{T} = \begin{bmatrix} 1 & 0 \\ 0 & e^{\frac{\pi}{4}i} \end{bmatrix}. \tag{7.9}$$

These two operations can be composed to approximate unitary transformations on a single qubit, such as the Pauli-X, Y, and Z gates (σ_x, σ_y, and σ_z) used to rotate the superposition along x-, y-, or z-axis. The Pauli gates are defined by

$$\sigma_x = \begin{bmatrix} 0 & 1 \\ 1 & 0 \end{bmatrix} = \mathbf{HT}^4\mathbf{H}, \tag{7.10}$$

$$\sigma_y = \begin{bmatrix} 0 & -i \\ i & 0 \end{bmatrix} = \mathbf{T}^2\mathbf{HT}^4\mathbf{HT}^6, \tag{7.11}$$

$$\sigma_z = \begin{bmatrix} 1 & 0 \\ 0 & -1 \end{bmatrix} = \mathbf{T}^4. \tag{7.12}$$

The Pauli matrices are involutory, that is, a matrix that is its own inverse, such that

$$\sigma_x^2 = \sigma_y^2 = \sigma_z^2 = -i\sigma_x\sigma_y\sigma_z = \begin{bmatrix} 1 & 0 \\ 0 & 1 \end{bmatrix} = \mathbf{I}. \tag{7.13}$$

It can be shown that by a given angle γ and a Pauli matrix $\sigma_{\mathbf{a}}, \mathbf{a} = [x, y, z]$

$$\exp(i\,\gamma\sigma_{\mathbf{a}}) = \cos(\gamma)\mathbf{I} + i\,\sin(\gamma)\sigma_{\mathbf{a}}. \tag{7.14}$$

Therefore, the rotations operator, which rotate the unit Bloch vector by an angle γ around a specific axis, is given by

$$R_x(\gamma) = e^{-i\frac{\gamma}{2}\sigma_x} = \cos\left(\frac{\gamma}{2}\right)\mathbf{I} - i\,\sin\left(\frac{\gamma}{2}\right)\sigma_x, \tag{7.15}$$

$$R_y(\gamma) = e^{-i\frac{\gamma}{2}\sigma_y} = \cos\left(\frac{\gamma}{2}\right)\mathbf{I} - i\,\sin\left(\frac{\gamma}{2}\right)\sigma_y, \tag{7.16}$$

$$R_z(\gamma) = e^{-i\frac{\gamma}{2}\sigma_z} = \cos\left(\frac{\gamma}{2}\right)\mathbf{I} - i\,\sin\left(\frac{\gamma}{2}\right)\sigma_z. \tag{7.17}$$

Moreover, the rotation operators can be expanded as

$$R_x(\gamma) = \begin{bmatrix} \cos(\gamma/2) & -i\sin(\gamma/2) \\ -i\sin(\gamma/2) & \cos(\gamma/2) \end{bmatrix}, \tag{7.18}$$

$$R_y(\gamma) = \begin{bmatrix} \cos(\gamma/2) & -\sin(\gamma/2) \\ \sin(\gamma/2) & \cos(\gamma/2) \end{bmatrix}, \tag{7.19}$$

$$R_z(\gamma) = \begin{bmatrix} e^{-\frac{\gamma}{2}i} & 0 \\ 0 & e^{\frac{\gamma}{2}i} \end{bmatrix}. \tag{7.20}$$

A sequence of quantum operations applied to the state of a given system is described by *quantum circuits*. Quantum circuits can be structured or random. In the first case, a structured quantum circuit is a sequence of well-defined quantum gates. On the other hand, the set of quantum gates are randomly selected. A circuit has a depth d equals the total number of layers conforming to the circuit. A layer is a circuit whose gates act on disjoint qubits, that is, a layer has a depth of one.

Example 3. *For instance, let us consider the following quantum circuit that rotates a qubit in the x-axis and afterward around the y-axis. First, a qubit in the ground state $|0\rangle = [1,0]^\top$ is rotated through the angle γ_1 around the x-axis applying the gate $R_x(\gamma_1)$, and eventually, the y-axis by the angle γ_2 via the gate $R_y(\gamma_2)$. After these operations, the qubit is now in the state*

$$|\psi\rangle = R_y(\gamma_2)R_x(\gamma_1)|0\rangle.$$

7.3.1.3 Qubit Measurements

A measurement must extract the information of a qubit, yielding a numerical result. If a quantum state $|\psi\rangle$ is measured, then only one classical state $|j\rangle$ can be seen with probability $|\alpha_j|^2$, which corresponds to the amplitude α_j. This is known as Born's rule. Accordingly, observing a quantum state induces a measurement in the computational basis state probabilities. However, other kinds of measurements can be computed from quantum systems, such as the expectation of an *observable* and its variance. Observables are Hermitian matrix M such that their eigenvalues are real numbers. The Pauli gates (σ_x, σ_y, and σ_z) are examples of observables, with eigenvalues ± 1. Ergo, the expected value of an observable M on a state $|\psi\rangle$ is

$$\langle M \rangle_\psi = \langle \psi | M | \psi \rangle. \tag{7.21}$$

Example 4. *Let us consider the state $|\psi\rangle = R_y(\gamma_2)R_x(\gamma_1)|0\rangle$, applying a measurement over the Pauli-Z observable, the expectation value is given by $\langle \sigma_z \rangle_\psi = \langle \psi | \sigma_z | \psi \rangle$. Depending on the circuit parameters γ_1 and γ_2, the output is yielded in the range $[-1, 1]$. For instance, if $\gamma_1 = 0.5$ and $\gamma_2 = 1.0$ then $\langle \sigma_z \rangle_\psi = 0.47416$.*

7.3.2 CONVOLUTIONAL NEURAL NETWORKS

Convolutional Neural Networks (CNNs) consist of a finite number of convolutional layers that can accurately extract intricate visual features such as edges, corners, interest points, among others, through a set of input images [2]. It can be applied to many practical problems such as detection, recognition, and classification. CNNs are a specific type of neural networks composed of the following layers: convolution layer, pooling layer, and fully connected layer. The convolution layer and the pooling layer can be fine-tuned concerning the hyper-parameters described in the next sections. Figure 7.4 illustrates the typical architecture of a CNN.

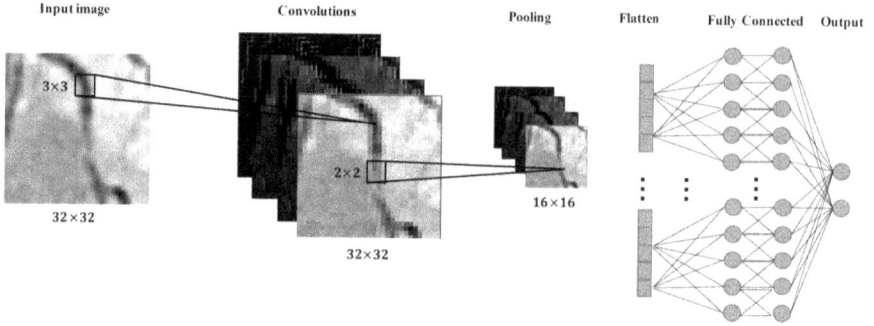

Figure 7.4: Typical architecture of a Convolutional Neural Network. The design comprises one convolutional layer with a kernel size of 3×3, followed by a pooling layer with a window size of 2×2. Finally, a fully connected layer is involved into the model used to optimize the objective functions.

7.3.2.1 Convolutional Layer

The convolutional layer uses filters to perform a set of convolution operations across the input image, concerning diverse filter dimensions. These layers can be viewed as 2-D kernels (matrices) where their hyper-parameters include the kernel size F and stride S. The stride S denotes the number of pixels by which the kernel moves after each operation. Hence, padding is introduced to enlarge the input such that the output has the same height/width dimension as the input. The resulting outcome **O** is called a *feature map* or *activation map*. Thus, a convolutional layer consists of K kernels trained to detect local features. Each feature map $\mathbf{O}_k$ is defined as follows:

$$\mathbf{O}_k = f\left(b_k + \sum_c \mathbf{w}_k[c] * \mathbf{I}[c]\right), \tag{7.22}$$

where f is an activation function (e.g., Sigmoid), b_k is the k-bias, $\mathbf{I}[c]$ is the image at the c-channel, $\mathbf{w}_k[c]$ is the k-kernel for such a channel, and $*$ denotes the convolution operator.

Example 5. *Consider a 5×5 input image represented as a matrix; also consider another 3×3 matrix representing a kernel. Here, the kernel will convolve over each pixel position of the input image to finally get a feature map, as shown in Figure 7.5. In this convolution layer, the hyper-parameters $F = 3 \times 3$, $S = 1$, and zero padding were used.*

7.3.2.2 Pooling Layer

The pooling layer, also called the subsampling or downsampling layer, typically applied after a convolutional layer, reduces each feature map's spatial size but retains the most valuable information. Pooling layers reduce the number of parameters and computations in the network and control overfitting [25]. Diverse types of pooling

Input

0	0	0	0	0	0	0
0	0	1	1	1	0	0
0	0	0	1	1	1	0
0	0	0	0	1	1	0
0	0	0	0	1	1	0
0	0	0	2	1	2	0
0	0	0	0	0	0	0

Kernel

1	1	-1
0	1	0
0	-1	-1

Output

0	2	3	2	0
0	-1	0	1	0
0	0	0	1	-1
0	2	3	1	0
0	0	2	0	0

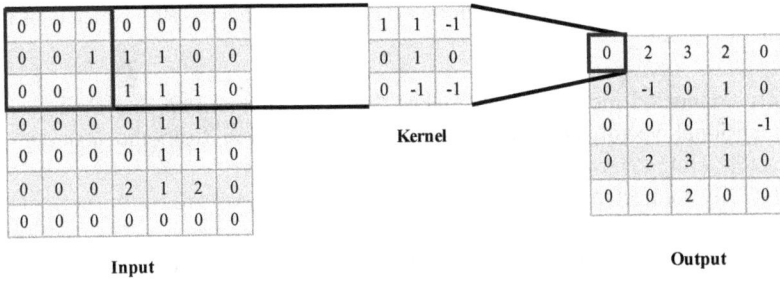

Figure 7.5: Convolution operation configured with zero-padding, a stride of 1 pixel, and a kernel size of 3×3.

can be applied, the max and average pooling are the most used. In max-pooling, the maximum value is taken from a certain spatial neighborhood (for example, a 2×2 window). Instead of taking the largest element, the average-pooling selects the average value of the neighborhood.

Example 6. *Given a feature map of size 5×5 produced by a convolutional layer, a pooling layer is defined with a spatial neighborhood of 2×2. The pooling procedure is illustrated in Figure 7.6, where the max and average pooling are shown in detail.*

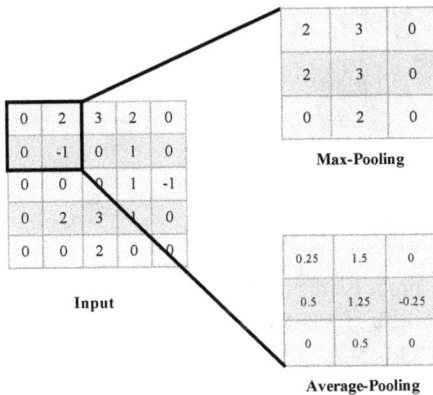

Input

0	2	3	2	0
0	-1	0	1	0
0	0	0	1	-1
0	2	3	1	0
0	0	2	0	0

Max-Pooling

2	3	0
2	3	0
0	2	0

Average-Pooling

0.25	1.5	0
0.5	1.25	-0.25
0	0.5	0

Figure 7.6: Pooling operations with a neighborhood of 2×2 and a stride of 2.

7.3.2.3 Fully Connected Layer

The fully connected layer takes the result of the convolution/pooling process to be employed for classification. The previous convolutional or pooling layer output is flattened into a single vector, each indexed value is representing a probability that a specific feature belongs to a label. Fully connected layers are usually found at the end of the CNN architectures; each neuron in the past layer is associated with each neuron on the next layer. In summary, the fully connected step consists of three layers:

Fully connected input layer: It makes the last features map into a single one-dimensional vector, commonly used in the transition from the last convolutional/pooling layer to the fully connected layer.

First fully connected layer: A set of weights are learned to predict the correct label from one-dimensional feature representation.

Fully connected output layer: It retrieves the final probabilities for each class.

7.3.2.4 Activation Functions

Activation functions are mathematical gates that determine the neuron's output transmitted to the next layer. Some activation functions normalize each neuron output into a range of $[1,0]$ or $[-1,1]$. Modern neural networks use non-linear activation functions to learn a complex representation between the network's inputs and outputs, as shown in Figure 7.7. The suitable choice of the activation function depends on

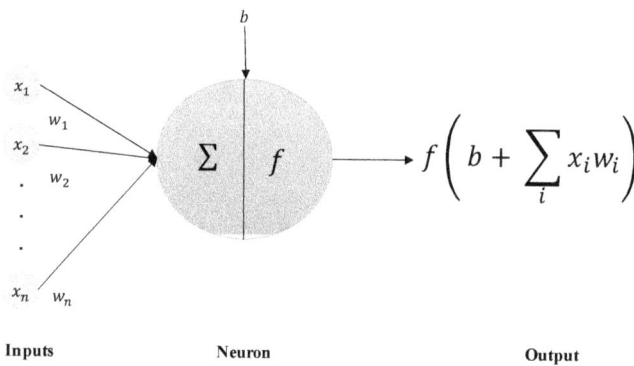

Inputs   **Neuron**   **Output**

Figure 7.7: The basic process of an activation function. The sum of all weighted inputs $(\mathbf{w}^\mathsf{T}\mathbf{x})$ and the bias $\mathbf{b}$ is passed through a non-linear activation function f to generate the neuron output.

the nature of the problem to solve. The most widely used activation functions are the Sigmoid, Hyperbolic Tangent, Rectified Linear Unit (ReLU), Leaky ReLU, and SoftMax. The Sigmoid function returns a value close to zero for small values in the argument and close to 1 for large argument values,

$$\text{Sigmoid}(x) = \frac{1}{1+e^{-x}}. \tag{7.23}$$

As an alternative to the Sigmoid function, the Hyperbolic Tangent function could be used as an activation function producing outputs in the range of $[-1,1]$ and is formally defined as

$$\tanh(x) = \frac{e^x - e^{-x}}{e^x + e^{-x}}. \qquad (7.24)$$

The Rectified Linear Unit (ReLU) activation function returns the element-wise maximum between 0 and the input value as follows:

$$\text{ReLU}(x) = \max(0, x). \qquad (7.25)$$

Leaky ReLU is a variation of the ReLU function, which has a small positive value when the input is not active

$$\text{Leaky-ReLU}(x) = \max(\delta x, x). \qquad (7.26)$$

Lastly, the SoftMax function converts a real input vector into a vector of probabilities. Therefore, the elements of the output vector must sum up to 1. The SoftMax function applied on the vector $\mathbf{x}$ is computed as

$$\text{SoftMax}(\mathbf{x}) = \frac{e^{\mathbf{x}}}{\sum\limits_{i=1}^{N} e^{x_i}}. \qquad (7.27)$$

7.4   PROPOSED METHOD

In the presented work, a *Quantum Convolutional Layer* (QCL) was employed as a new type of image preprocessing method. From a given XCA image, the QCL generates a multichannel image resembling the feature maps generated by a typical Convolutional Layer. Next, the outcome of the QCL feeds a traditional CNN focused on detecting atherosclerosis. A general framework of the proposed method is shown in Figure 7.8 and discussed in this section.

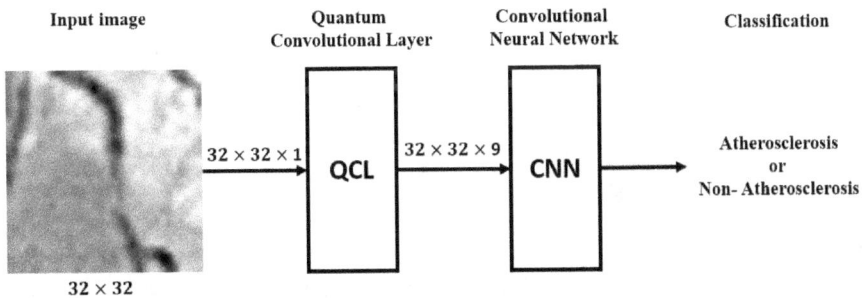

Figure 7.8: Outline of the proposed methodology. The input XCA patches pass through a QCL to generate a 9-channel image to feed the traditional CNN to detect atherosclerosis.

7.4.1 QUANTUM CONVOLUTIONAL LAYER

In traditional image processing methods, there are several techniques to enhance the image quality to accelerate and improve the training process in machine learning algorithms. Moreover, new promising hybrid DL methods have emerged. The QCL is an example of this approach. A QCL produces an advanced set of feature maps when applied to an input image. However, unlike the classical convolution operation where each filter produces a feature map, a quantum convolution employs a quantum circuit as a filter, and it can generate from each circuit multiple feature maps. A QCL requires some hyper-parameters to be tuned: the number of filters and their respective input size. Besides, for each filter, an encoding function is necessary to initialize the quantum state. Finally, the feature maps are generated from a given decoding function (an observable). It is noteworthy that the filter size defines the number of qubits required to operate the quantum circuit. In this study, a single 3×3 filter size was chosen. Therefore, the circuits dispose of nine qubits and compute nine feature maps. The proposed methodology is described in the following procedure and illustrated in Figure 7.10.

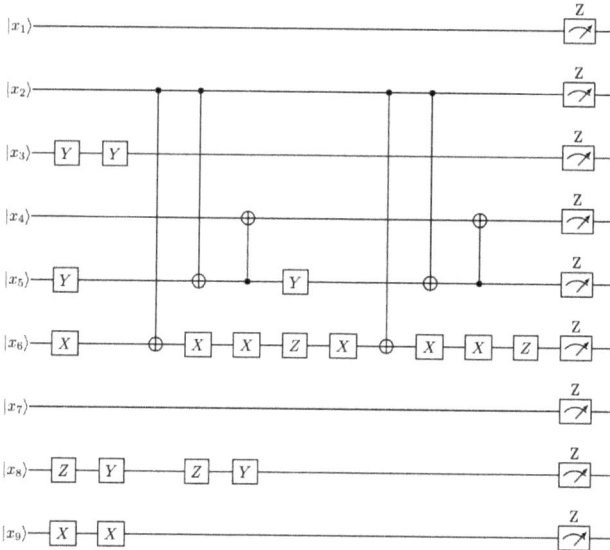

Figure 7.9: 9-qubit Random Quantum Circuit employed in the Quantum Convolutional Layer. The circuit has the Pauli-X, Pauli-Y, and Pauli-Z rotations randomly distributed and controlled-CNOT gates as imprimitives (2-qubit entangled gates).

1. Ground state initialization: The input image sub-regions **u** were obtained by a sliding window of 3×3 pixels and a stride of 1 pixel to initialize the quantum ground state as $|x_n\rangle = R_y(u_n)$. Notice that the ground state is parameterized as

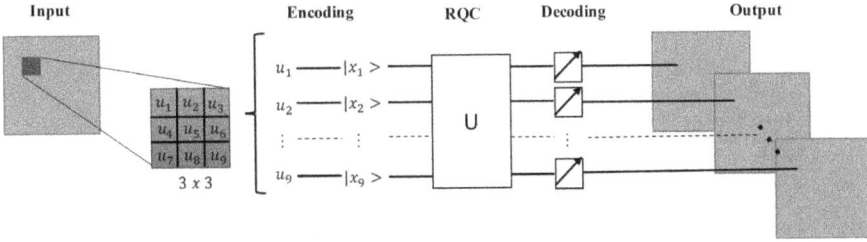

Figure 7.10: Design of the Quantum Convolutional Layer consisting of three main stages: ground-state initialization (encoding), Random Quantum Circuit (RQC) definition and output state computation, and the measurement (decoding). From a sliding window of 3×3, the QCL generates nine expectation values mapped into nine different channels from data.

rotation over the y-axis as encoding function, with an angle $\gamma = u_n$, where u_n is the intensity of the pixel n in the image neighborhood $\mathbf{u}$.

2. Random quantum circuit definition: A random quantum circuit, associated with a unitary $\mathbf{U}$, is generated. Hence, the circuit comprises layers of randomly chosen single-qubit rotations and 2-qubit entangling gates. The 2-qubit gates and the rotations are randomly distributed in the circuit. Specifically, a 9-qubit circuit with two layers is generated, as shown in Figure 7.9.

3. Output state computation: Each image neighborhood $\mathbf{u}$ is processed by the quantum circuit, whereby $|x'\rangle = \mathbf{U}|x\rangle$

4. Measurement: The system is measure using a Pauli-Z gate, observable over the respective output state, such that $\langle \sigma_z \rangle_{x'} = \langle x' | \sigma_z | x' \rangle$. Each nine expectation values are mapped into nine channels of a single output pixel.

7.4.2   NETWORK ARCHITECTURE

Two CNN architectures were tested to accurately detect atherosclerosis. For each CNN, the detection performance was carried out using the images generated by the QCL and the original XCA images. Table 7.1 presents a detailed description of these two architectures. The first network introduced by Au et al. [8] is based on a DenseNet architecture [9]. On the other hand, the second network proposed by Antczak and Liberadzki [10] is based on a VGG architecture [11] but changing the kernel size and depth as well as removing the pooling layers.

The CNN weights were trained from scratch, with a random initialization using the Glorot (also known as Xavier) algorithm [26]. The fundamental idea is to initialize each layer weight with a small Gaussian value characterized by a zero mean and variance based on the number of columns (n) in the previous layer $\mathbf{w}^{k-1}$. In consequence, the weights are scaled by the inverse of the square root of n as follows:

$$\mathbf{w}^k \sim \mathcal{N}\left(-\frac{1}{\sqrt{n}}, \frac{1}{\sqrt{n}}\right). \tag{7.28}$$

Table 7.1

CNN Architectures for Atherosclerosis Detection

Network Architecture	Layer	Description
	64-Conv(3 × 3)	Convolutional layer with 64 kernels, each with a size of 3 × 3 pixels
	64-Conv(1 × 1)	Convolutional layer with 64 kernels, each with a size of 1 × 1 pixels
	64-Conv(3 × 3)	Convolutional layer with 64 kernels, each with a size of 3 × 3 pixels
I) Au *et al.* [8]	Conc(1, 3)	Concatenation of the outputs of the first and third convolutional layer
	64-Conv(1 × 1)	Convolutional layer with 64 kernels, each with a size of 1 × 1 pixels
	64-Conv(3 × 3)	Convolutional layer with 64 kernels, each with a size of 3 × 3 pixels
	Conc(1, 6)	Concatenation of the outputs of the first and sixth convolutional layer
	GMP	Global Max Pooling layer
	1-Dense	Dense layer with one neuron and a Sigmoid activation function
	8-Conv(7 × 7)	Convolutional layer with eight kernels, each with a size of 7 × 7 pixels
	Dropout(0.5)	Dropout layer with a rate of 0.5
	8-Conv(7 × 7)	Convolutional layer with eight kernels, each with a size of 7 × 7 pixels
II) Antczak and Liberadzki [10]	8-Conv(7 × 7)	Convolutional layer with eight kernels, each with a size of 7 × 7 pixels
	Dropout(0.5)	Dropout layer with a rate of 0.5
	8-Conv(7 × 7)	Convolutional layer with eight kernels, each with a size of 7 × 7 pixels
	Dropout(0.5)	Dropout layer with a rate of 0.5
	16-Dense	Dense layer with 16 neurons and a Sigmoid activation function
	1-Dense	Dense layer with one neuron and a Sigmoid activation function

Let y_n be the class indicator variable for the n-th input patch, which operates as

$$y_n = \begin{cases} 0, & \text{if non-atherosclerosis,} \\ 1, & \text{if atherosclerosis.} \end{cases} \qquad (7.29)$$

Furthermore, $\hat{y}$ corresponds to the estimated probability by the CNN that the input patch has atherosclerosis. Lastly, the optimization process is achieved by minimizing the binary cross-entropy loss function defined as

$$\min_{\mathbf{w}} J(y,\hat{y}) = -\frac{1}{N}\sum_{n=1}^{N}[y_n log(\hat{y}_n) + (1-y_n)log(1-\hat{y}_n)], \qquad (7.30)$$

where N represents the batch-size used during the minimization. For artherosclerosis detection, the loss function can be rewritten as

$$J(y_n,\hat{y}_n) = -\begin{cases} log(\hat{y}_n), & \text{if } y_n = 1, \\ log(1-\hat{y}_n), & \text{if } y_n = 0, \end{cases} \qquad (7.31)$$

where $y = 1$ means that the class $C_1 = C_i$ is a positive case of atherosclerosis. The gradient computed with respect to each predicted label $\hat{y}_n$ will only depend on the loss given by its binary problem.

The objective function from (7.31) is minimized through the training step by computing its gradient regarding the current weights' values. In most neural networks, the Stochastic Gradient Descent (SGD) optimization [27] is employed to find an optimal solution. SGD computes the gradient of the parameters using only a few training samples (batch size). For SGD optimization, the gradient is updated by

$$\mathbf{w}_{k+1} = \mathbf{w}_k - \alpha\nabla_{\mathbf{w}_k}J(y,\hat{y}), \qquad (7.32)$$

where α is the learning rate. SGD can lead to slow convergence, particularly after the initial steep gains. Some methods have been incorporated to overcome such inconvenience. Thus, the momentum is a method that integrates the past gradients in each dimension. In SGD with momentum [28] (SGDM), the gradient at every dimension is incorporated to gain velocity where the parameters have a consistent gradient. The momentum update is given by

$$v_{k+1} = \eta v_k + \alpha\nabla_{\mathbf{w}_k}J(y,\hat{y}),$$
$$\mathbf{w}_{k+1} = \mathbf{w}_k - v_{k+1} \qquad (7.33)$$

where v is the velocity term, starting with the initial value zero, and η is the momentum coefficient that controls the fraction of the previous velocity term to be considered in the current update step. The evaluated networks were optimized using SGD and SGDM with a learning rate $\alpha = 0.01$ and momentum $\eta = 0.8$, respectively, with a batch size $N = 100$ during $1,000$ epochs.

7.4.3 EVALUATION METRICS

A set of binary classification metrics were considered to evaluate the network's performance in the detection of atherosclerosis. These metrics include the Accuracy, Sensitivity, Specificity, Precision, and F-Measure.

Accuracy refers to the ratio of correct predictions over the number of instances evaluated,

$$\text{Accuracy} = \frac{t_p + t_n}{t_p + f_p + t_n + f_n}, \qquad (7.34)$$

where t_p refers to the true positive cases of atherosclerosis detection, t_n refers to the true negative cases, f_p represents the number of instances incorrectly classified as positive (false positives), and f_n represents the number of instances incorrectly classified as negative (false negatives).

Sensitivity measures the fraction of positive instances correctly classified, which is given by

$$\text{Sensitivity} = \frac{t_p}{t_p + f_n}. \tag{7.35}$$

Specificity measures the proportion of negatives samples correctly classified as follows

$$\text{Specificity} = \frac{t_n}{t_n + f_p}. \tag{7.36}$$

Precision is used to measure the positive patterns correctly classified from the total predicted patterns in a positive class. It is calculated as

$$\text{Precision} = \frac{t_p}{t_p + f_p}. \tag{7.37}$$

F-Measure represents the harmonic mean between recall (sensitivity) and precision values. A general F-Measure is described as follows

$$F_\beta\text{-Measure} = (1 + \beta^2) \times \frac{\text{Precision} \times \text{Recall}}{\beta^2 \times \text{Precision} + \text{Recall}}. \tag{7.38}$$

In particular, if $\beta = 1$, a balanced F_β-Measure is obtained; consequently, such expression is acknowledged as the F_1-Measure.

7.5 RESULTS AND DISCUSSIONS

The atherosclerosis detection was evaluated using two different approaches to train CNN. First, a quantum convolutional layer performs a preprocessing over the XCA images. On the other hand, CNN is trained using the original XCA images. The detection results for each strategy are presented. Moreover, a comparative analysis of the proposed approach using SGD and SGDM is included in this section. The analysis concerned the CNN architectures introduced by Au *et al.* [8] and Antczak and Liberadzki [10].

For the computational experiments, Python 3.6, Keras 2.3.1, TensorFlow 2.2.0, and the PennyLane library [19] were used and tested on a Cloud Platform, including CPU and GPU processors. The CPU included an Intel (R) Xeon (R), 12 GB of RAM, and a 2.00 GHz dual processor. The GPU was powered by a Tesla P4 consisting of 2560 CUDA cores and 8 GB VRAM.

7.5.1 DATASET OF CORONARY STENOSIS

Antczak and Liberadzki [10] introduced a dataset for stenosis detection, publicly available on the authors GitHub site [29]. The dataset consists of 250 patches of size 32×32 pixels obtained from XCA images. In numerical experiments, the pixel values were normalized into $[0, 1]$.

The patches were obtained, as seen in Figure 7.11, following the next steps:

1. Input images are downsampled from 256×256 to 128×128 pixels.
2. A sliding window of size 32×32 was moved to obtain a set of patches from each image.
3. Each patch is labeled as non-stenosis, stenosis, respectively.

The sliding window produces multiple overlapping matches; thus, multiple patches may be classified as positive cases, even if there is only one stenosis on the image.

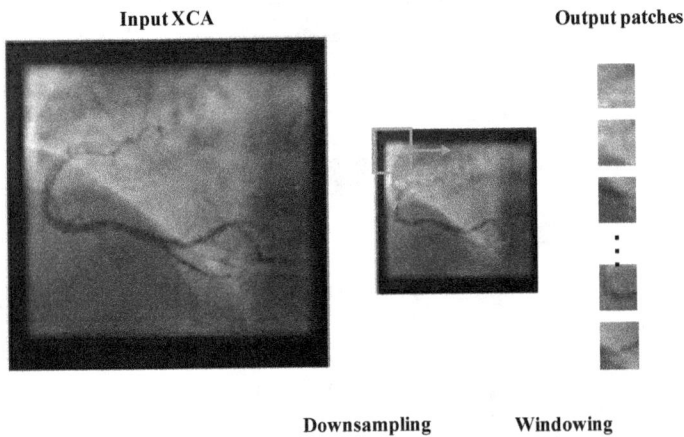

Figure 7.11: General outline of XCA patches generation. First, the XCA image is sub-sampled; next, a sliding window generates the output patches.

7.5.2 QUANTUM PREPROCESSING

The quantum convolutional layer (QCL) acts as an image preprocessing. Given an XCA patch of 32×32 for one channel, the QCL outcome is a patch of 32×32 with nine channels. The input pixel values are in the range of $[0, 1]$, and the output is in the range $[-1, 1]$.

Figure 7.12 shows the effect of the quantum convolution layer on a batch of samples from XCA images. The QCL results suggest that this layer generates different images where the contrast is changing in each channel. In other cases, in the fifth channel (sixth row), the QCL outcome is similar to extract the image texture (e.g., Local Binary Pattern [30]).

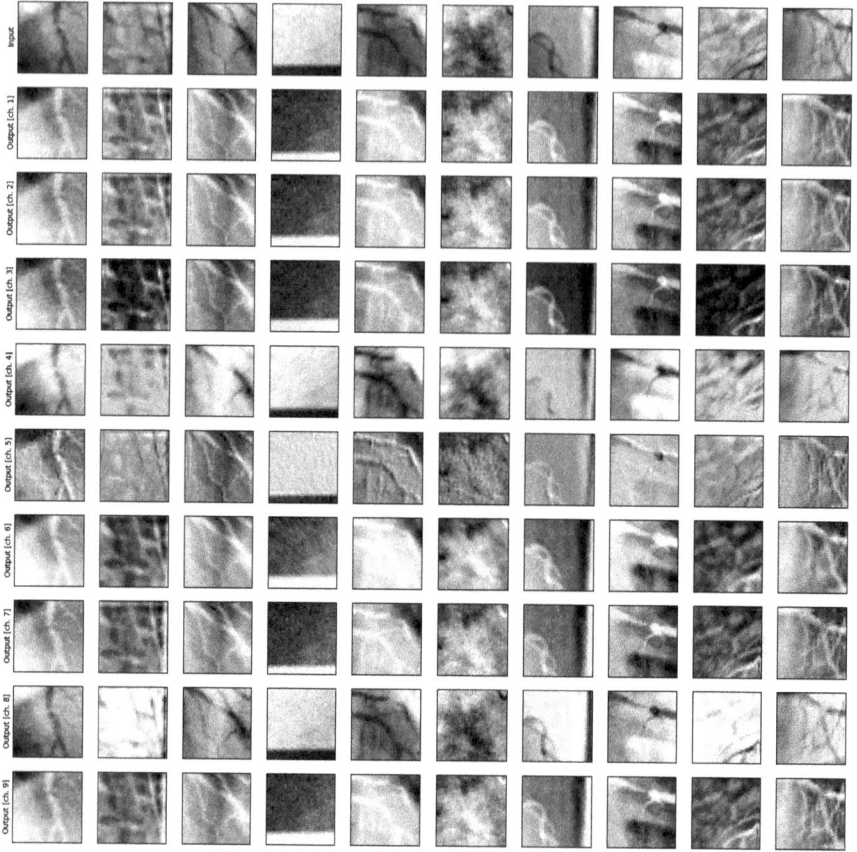

Figure 7.12: Feature maps generated by the Quantum Convolutional Layer (QCL). First row: raw XCA images. The following nine rows show the generated images by the QCL.

7.5.3   TRAINING RESULTS

A k-Fold cross-validation procedure was carried out during the numerical experiments. In k-Fold cross-validation, a parameter k decides in how many folds the dataset is divided. Thus, each fold gets the chance to appear in the training set $k-1$ times. The number of folds chosen was $k = 5$, as illustrated in Figure 7.13.

The dataset was divided into two subsets: training and testing of 125 images each. The testing subset was excluded during the cross-validation process. In this approach, four folds (75% of the training subset) were used to train the CNN, and the remaining fold (25% of the training subset) was saved for validation. This process was executed five times, selecting the best model for validation loss in each iteration. The best model was updated if the validation loss is improved over the epochs; this avoids overfitting. Finally, each iteration's atherosclerosis detection performance was evaluated using the excluded initial subset (testing). Furthermore, the mean and maximum coefficients were obtained, as well as the standard deviation for these coefficients.

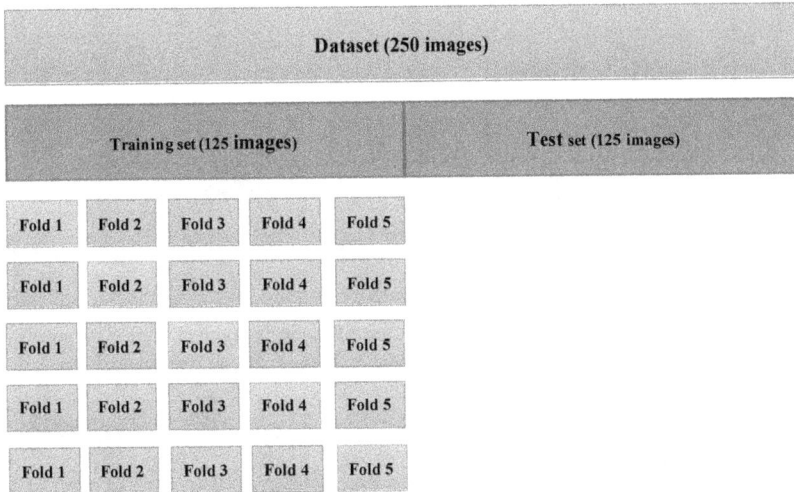

Figure 7.13: k-Fold cross-validation procedure. The dataset was split into training and testing sets. Next, the training set is divided into folds to train CNN. The fold marked as gray was taken as validation, while the folds in light green as training data.

The learning curve can measure the model's performance during the training and validation steps in terms of accuracy and loss. In Figure 7.14, losses and accuracy curves are plotted concerning the number of epochs used to train the model.

Figure 7.15 shows that the accuracy (using SGDM) increases while the loss decreases. The Q-CNN-A (CNN proposed by Au *et al.* [8] with the quantum convolutional layer) achieves the best accuracy in training and validation. Those results are followed by the Q-CNN-B (CNN proposed by Antczak and Liberadzki [10] using the quantum convolutional layer). In terms of validation loss, the models start to overfit during the training process.

On the other hand, the accuracy curves in Figure 7.16 also shown that the CNN trained with the outcome of the QCL reach the highest accuracy (in particular, Q-CNN-A) by using SGD as the optimizer. This same network architecture but using only the raw (normalized) XCA for training, followed in accuracy performance.

In contrast to the loss curves obtained by SGDM, the curves in Figure 7.17 suggest that the validation loss decreases and has a small gap with the training loss during the training process.

7.5.4 DETECTION RESULTS

In this research, the approach analysis was carried out using the generated XCA patches by the QCL and the original (normalized) XCA patches. Additionally, the

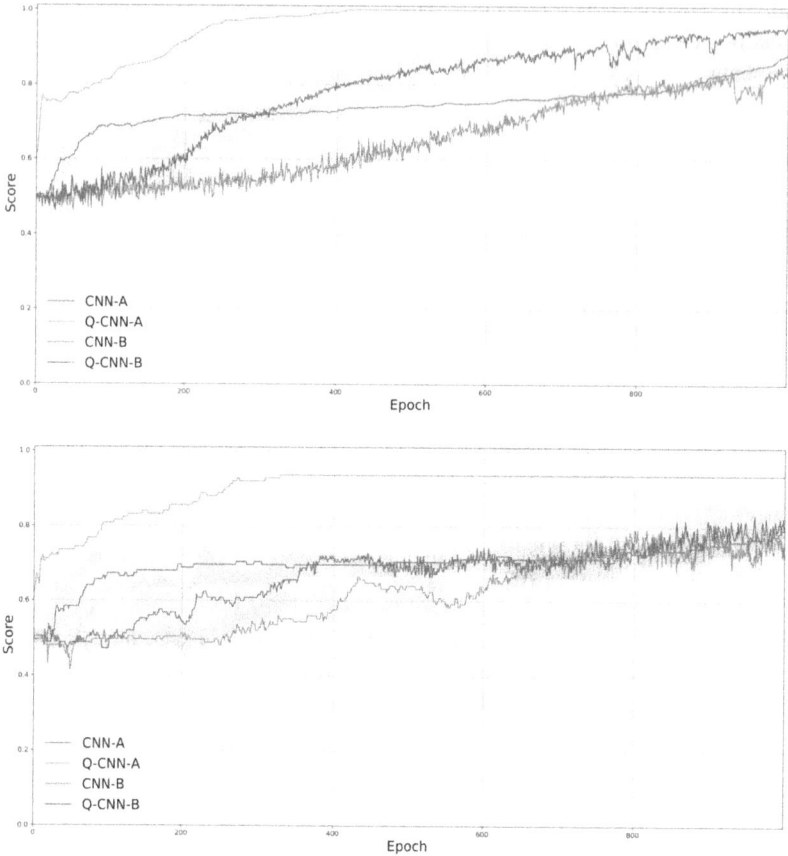

Figure 7.14: Training and validation accuracy curves of the CNN models employing SGD with Momentum concerning the number of epochs (1000) in the training phase. Top: Training accuracy curves, Bottom: Validation accuracy curves.

detection results were compared using SGD and the innovative variant using momentum (SGDM). Table 7.2 presents the performance rates of atherosclerosis detection for each CNN studied. It shows that the use of momentum provides improved performance for both CNN architectures. The best performance for each metric was highlighted.

Furthermore, the network performance surpassed the raw-XCA patches version when the network is trained using the QCL preprocessing, reaching an accuracy of 83.36% against 79.84% for the architecture employed by Au *et al.* [8].

For the second architecture (Antczak and Liberadzki [10]), the accuracy is improved from 69.44% to 80.80% when the network is fed with the QCL output images. In general, an improvement for precision, sensitivity, specificity, and F_1 score was achieved using the QCL.

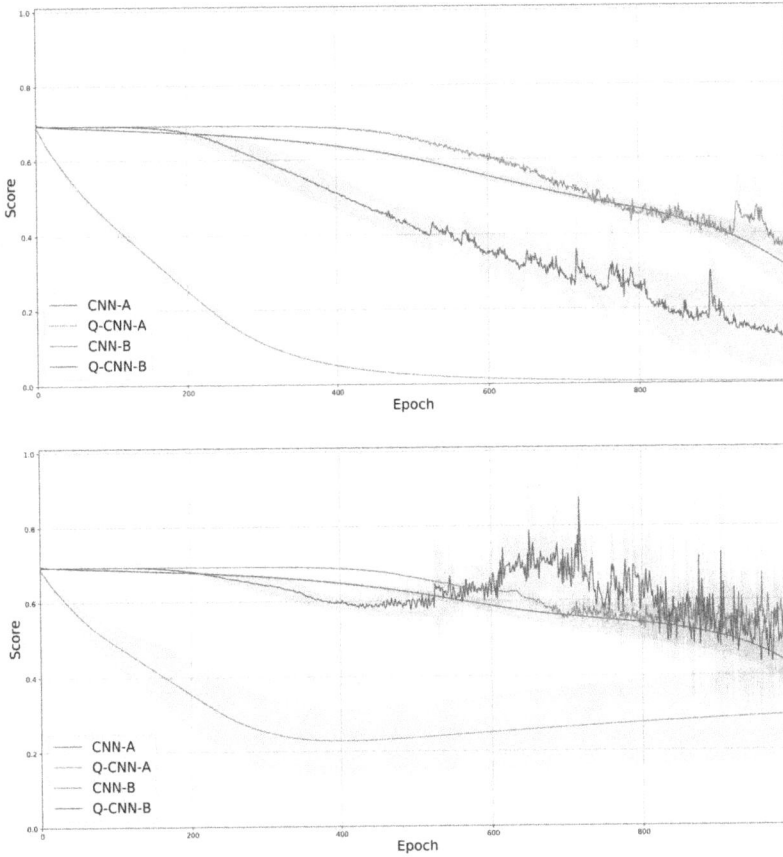

Figure 7.15: Training and validation loss curves of the CNN models employing SGD with Momentum concerning the number of epochs (1000) in the training phase. Top: Training loss curves, Bottom: Validation loss curves.

Figure 7.18 shows a prediction sample: true-positive, true-negative, false-positive, and false-negative cases. Beneath each case, the generated multichannel image by the QCL was presented. Therefore, in the case of correct detection (true-positive and true-negative), the vessel pixels were enhanced in each channel of the QCL image outcome.

Meanwhile, misclassification cases were also found. Thus, in the false-positive case, the network detects as atherosclerosis a patch with no vessel pixels, containing a central region with noise that looks like a vessel portion. In the false-negative samples, the atherosclerosis region contains a cut in the artery, which causes that vessel pixels to have background intensities both in the original image and in the multichannel image generated by the QCL.

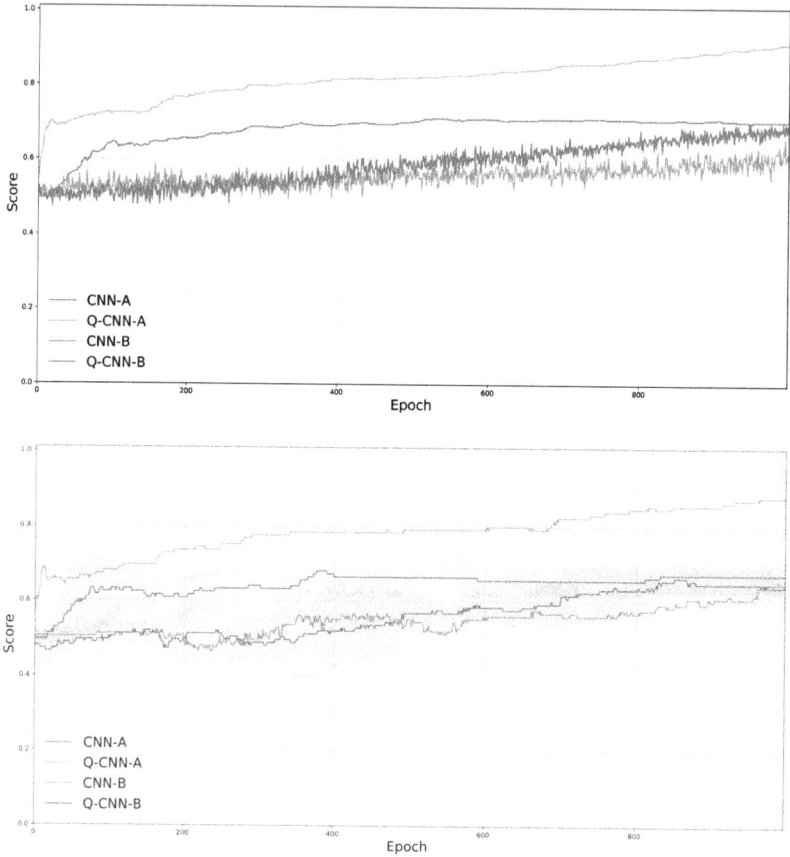

Figure 7.16: Training and validation accuracy curves of the CNN models employing SGD, and concerning as 1000 the number of epochs in the training phase. Top: Training accuracy curves, Bottom: Validation accuracy curves.

7.6   CONCLUDING REMARKS

In this study, a Hybrid Quantum-Convolutional Neural Network method for atherosclerosis detection in XCA images was introduced. The proposed scheme includes a Quantum Convolutional Layer used as a preprocessing to improve the atherosclerosis detection performance of typical CNNs. The numerical experiments, based on two distinct CNN architectures (DenseNet-based and VGG-based architectures) and trying on two different training algorithms: Stochastic Gradient Descent and Stochastic Gradient Descent with Momentum, have demonstrated that using a Quantum Convolutional Layer on a limited XCA dataset performs efficiently for atherosclerosis detection. The introduced hybrid methodology improved

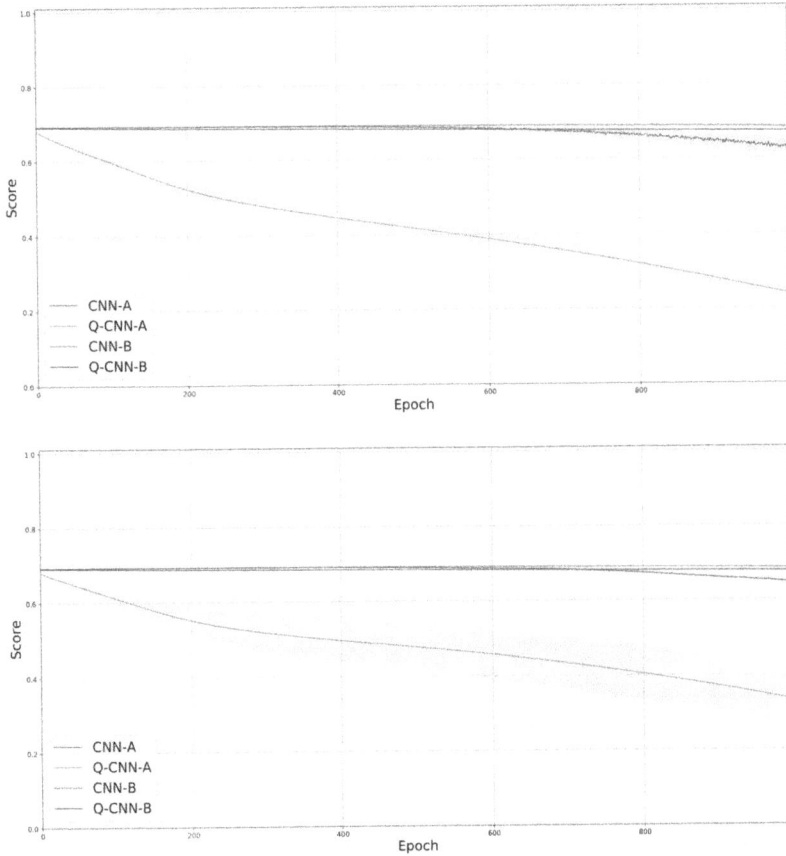

Figure 7.17: Training and validation loss curves of the CNN models employing SGD concerning the number of epochs (1000) in the training phase. Top: Training loss curves, Bottom: Validation loss curves.

the five evaluation metrics concerning the classical CNN architectures trained with the normalized-raw XCA images.

Additionally, the numerical results are permitted to assert that momentum during the optimization step improves the detection performance. The atherosclerosis detection with the DenseNet-based CNN employing the Quantum Convolutional Layer reached the best accuracy, with an 83.36%, while in precision, sensitivity, specificity, and F1-score achieved an 81.52%, 86.67%, 83.99%, and 80.00%, respectively.

A future direction of this work will incorporate this quantum layer during the optimization procedure. Finally, in agreement with numerical results, the proposed method has shown the potential of a preprocessing Quantum Convolutional Layer to generate discriminative feature maps (a multichannel image) to feed a classical CNN to detect atherosclerosis in XCA images.

Table 7.2

Network Detection Results.

Model	Accuracy	Precision	Sensitivity	F_1-Score	Specificity
QCL-SGDM-A	**83.36%** (±**2.11%**)	**81.52%** (±**2.13%**)	86.67% (±2.77%)	83.99% (±2.07%)	**80.00%** (±**2.62%**)
QCL-SGDM-B	80.80% (±3.16%)	78.55% (±2.16%)	85.08% (±5.37%)	81.64% (±3.41%)	76.45% (±2.19%)
QCL-SGD-A	82.88% (±1.65%)	78.46% (±1.94%)	**91.11%** (±**1.90%**)	**84.29%** (±**1.42%**)	74.52% (±2.96%)
QCL-SGD-B	60.64% (±1.55%)	59.09% (±1.16%)	71.75% (±9.81%)	64.47% (±3.72%)	49.35% (±8.20%)
SGDM-A	79.84% (±4.28%)	75.98% (±3.15%)	87.62% (±6.06%)	81.35% (±4.26%)	71.94% (±3.32%)
SGDM-B	69.44% (±1.99%)	67.80% (±2.28%)	75.87% (±10.06%)	71.13% (±4.17%)	62.90% (±7.36%)
SGD-A	67.36% (±2.55%)	64.96% (±0.86%)	76.83% (±12.20%)	69.90% (±5.12%)	57.74% (±7.59%)
SGD-B	60.80% (±2.09%)	58.55% (±1.42%)	78.10% (±19.40%)	65.67% (±7.36%)	43.23% (±16.11%)

The detection performance for each neural network (A:Au *et al.* [8], B: Antczak and Liberadzki[10]) was carried out employing the Quantum Convolutional Layer (QCL) preprocessing and the original XCA. The networks were also optimized using Stochastic Gradient Descent (SGD) and SGD with Momentum (SGDM).

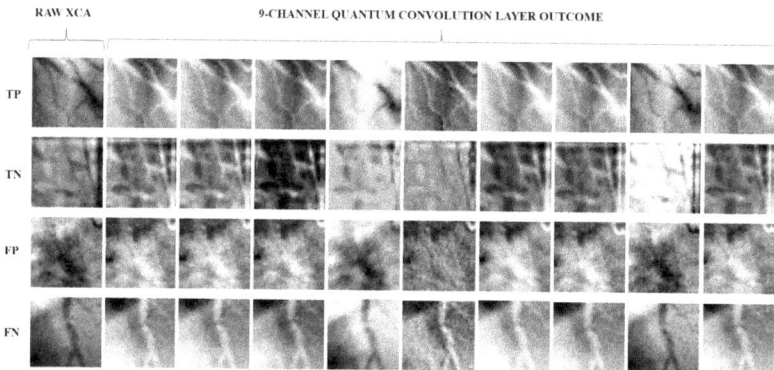

Figure 7.18: Detection result sample. Each row shows one case of True-Positive (TP), True-Negative (TN), False-Positive (FP), and False-Negative (FN). From left to right: one raw XCA images and the nine-channel QCL outcome image.

CONFLICTS OF INTEREST

The authors declare that there is no conflict of interest.

ACKNOWLEDGEMENTS

This research was supported by the Engineering Division of the Campus Irapuato-Guanajuato, grant NUA 147347; and the Mexican Council of Science and Technology CONACyT, Doctoral Studies Grants no. 626154 and 626155.

ETHICAL APPROVAL

All procedures performed in studies involving human participants were in accordance with the ethical standards of the institutional and/or national research committee and with the 1964 Helsinki Declaration and its later amendments or comparable ethical standards. For this type of study, formal consent is not required.

REFERENCES

1. World Health Organization. Cardiovascular Diseases (CVDs). https://www.who.int/news-room/fact-sheets/detail/cardiovascular-diseases-(cvds), Aug 2020.
2. Alex Krizhevsky, Ilya Sutskever, and Geoffrey E Hinton. Imagenet classification with deep convolutional neural networks. In Advances in Neural Information Processing Systems, pages 1097–1105, 2012.
3. Olga Russakovsky, Jia Deng, Hao Su, Jonathan Krause, Sanjeev Satheesh, Sean Ma, Zhiheng Huang, Andrej Karpathy, Aditya Khosla, Michael Bernstein, et al. Imagenet large scale visual recognition challenge. International Journal of Computer Vision, 115(3):211–252, 2015.
4. Jacob Biamonte, Peter Wittek, Nicola Pancotti, Patrick Rebentrost, Nathan Wiebe, and Seth Lloyd. Quantum machine learning. Nature, 549(7671):195–202, 2017.
5. Chenxin Sui, Zhuang Fu, Zeyu Fu, Yao Wang, Yu Zhuang, Rongli Xie, Yanna Zhao, Jun Zhang, and Jian Fei. A Novel method for vessel segmentation and automatic diagnosis of vascular stenosis. In 2019 IEEE International Conference on Robotics and Biomimetics (ROBIO), pages 918–923. IEEE, 2019.
6. WeiWu, Jingyang Zhang, Hongzhi Xie, Yu Zhao, Shuyang Zhang, and Lixu Gu. Automatic detection of coronary artery stenosis by convolutional neural network with temporal constraint. Computers in Biology and Medicine, 118:103657, 2020.
7. S. Jevitha, M. Dhanalakshmi, and Pradeep G Nayar. Analysis of left main coronary bifurcation angle to detect stenosis. In International Conference on Intelligent Systems Design and Applications, pages 627–639. Springer, 2018.
8. Benjamin Au, Uri Shaham, Sanket Dhruva, Georgios Bouras, Ecaterina Cristea, Andreas Coppi, Fred Warner, Shu-Xia Li, and Harlan Krumholz. Automated characterization of stenosis in invasive coronary angiography images with convolutional neural networks. arXiv preprint arXiv:1807.10597, 2018.
9. Gao Huang, Zhuang Liu, Laurens Van Der Maaten, and Kilian Q Weinberger. Densely connected convolutional networks. In Proceedings of the IEEE Conference on Computer Vision and Pattern Recognition, pages 4700–4708, 2017.
10. Karol Antczak and Lukasz Liberadzki. Stenosis detection with deep convolutional neural networks. In MATEC Web of Conferences, volume 210, page 04001. EDP Sciences, 2018.
11. Karen Simonyan and Andrew Zisserman. Very deep convolutional networks for large-scale image recognition. arXiv preprint arXiv:1409.1556, 2014.

12. Chao Cong, Yoko Kato, Henrique Doria Vasconcellos, Joao Lima, and Bharath Venkatesh. Automated stenosis detection and classification in X-ray angiography using deep neural network. In 2019 IEEE International Conference on Bioinformatics and Biomedicine (BIBM), pages 1301–1308. IEEE, 2019.

13. Christian Szegedy, Wei Liu, Yangqing Jia, Pierre Sermanet, Scott Reed, Dragomir Anguelov, Dumitru Erhan, Vincent Vanhoucke, and Andrew Rabinovich. Going deeper with convolutions. In Proceedings of the IEEE Conference on Computer Vision and Pattern Recognition, pages 1–9, 2015

14. Emmanuel Ovalle-Magallanes, Juan Gabriel Avina-Cervantes, Ivan Cruz–Aceves, and Jose Ruiz-Pinales. Transfer learning for stenosis detection in X-ray coronary angiography. Mathematics, 8(9):1510, 2020.

15. Kaiming He, Xiangyu Zhang, Shaoqing Ren, and Jian Sun. Deep residual learning for image recognition. In Proceedings of the IEEE Conference on Computer Vision and Pattern Recognition, pages 770–778, 2016.

16. Maxwell Henderson, Samriddhi Shakya, Shashindra Pradhan, and Tristan Cook. Quanvolutional neural networks: Powering image recognition with quantum circuits. Quantum Machine Intelligence, 2(1):1–9, 2020.

17. Jennifer Sleeman, John Dorband, and Milton Halem. A hybrid quantum enabled RBM advantage: convolutional autoencoders for quantum image compression and generative learning. In Quantum Information Science, Sensing, and Computation XII, volume 11391, page 113910B. International Society for Optics and Photonics, 2020.

18. Vijayasri Iyer, Bhargava Ganti, A.M. Hima Vyshnavi, P.K. Krishnan Namboori, and Sriram Iyer. Hybrid quantum computing based early detection of skin cancer. Journal of Interdisciplinary Mathematics, 23(2):347–355, 2020.

19. Ville Bergholm, Josh Izaac, Maria Schuld, Christian Gogolin, Carsten Blank, Keri McKiernan, and Nathan Killoran. Pennylane: Automatic differentiation of hybrid quantum-classical computations. arXiv preprint arXiv:1811.04968, 2018.

20. Aradh Bisarya, Shubham Kumar, Walid El Maouaki, Sabyasachi Mukhopadhyay, Bikash K Behera, Prasanta K Panigrahi, et al. Breast Cancer Detection Using Quantum Convolutional Neural Networks: A Demonstration on a Quantum Computer. medRxiv, 2020.

21. Michael A. Nielsen and Isaac L. Chuang. Quantum Computation and Quantum Information: 10th Anniversary Edition. Cambridge University Press, 2010.

22. Ronald De Wolf. Quantum computing: Lecture notes. arXiv preprint arXiv:1907.09415, 2019.

23. J. Robert Johansson, Paul D Nation, and Franco Nori. QuTiP: An open-source Python framework for the dynamics of open quantum systems. Computer Physics Communications, 183(8):1760–1772, 2012.

24. Albert Einstein, Boris Podolsky, and Nathan Rosen. Can quantum-mechanical description of physical reality be considered complete? Physical Review, 47(10):777, 1935.

25. Dominik Scherer, Andreas Muller, and Sven Behnke. Evaluation of pooling operations in convolutional architectures for object recognition. In International Conference on Artificial Neural Networks, pages 92–101. Springer, 2010.

26. Xavier Glorot and Yoshua Bengio. Understanding the difficulty of training deep feedforward neural networks. In Proceedings of the thirteenth international conference on artificial intelligence and statistics, pages 249–256, 2010.

27. Herbert Robbins and Sutton Monro. A stochastic approximation method. The Annals of Mathematical Statistics, pages 400–407, 1951.

28. Ning Qian. On the momentum term in gradient descent learning algorithms. Neural Networks, 12(1):145–151, 1999.

29. Karol Antczak and Lukasz Liberadzki. Deep Stenosis Detection Dataset. https://github.com/KarolAntczak/DeepStenosisDetection, Aug 2020.

30. Timo Ojala, Matti Pietikainen, and David Harwood. A comparative study of texture measures with classification based on featured distributions. Pattern Recognition, 29(1):51–59, 1996.

8 Multilevel Quantum Elephant Herd Algorithm for Automatic Clustering of Hyperspectral Images

Automatic clustering of hyperspectral images is a very strenuous task due to the presence of huge number of redundant bands and complexity to process them. In this work, two quantum versions of Elephant Herd Optimization algorithm are proposed for this purpose. The binary and ternary quantum logics used enhances the exploration and exploitation capability of the elephant herd optimization. These algorithms are compared to their classical counterpart. They are implemented on the Salinas dataset. The proposed *qutrit* based algorithm is found to converge faster and produce more robust results. The Xie-Beni Index is used as the fitness function. A few statistical tests like mean, standard deviation and Kruskal Wallis test are performed to establish the efficiency of the proposed algorithm. The F score is used to compare the segmented images using the optimal cluster numbers. The proposed algorithms are found to perform better in most of the cases.

8.1 INTRODUCTION

Hyperspectral image (HSI) processing has caught the attention of many researchers in the past decade. The development of powerful spectral cameras provided researchers with the tools to easily acquire them. HSI can provide extensive and meticulous information about the object or area which are captured by means of the spectral cameras [5]. The reflection and absorption capabilities of different materials present on the Earth's surface are unique. These spectral informations are used by HSI to recognize them individually [41]. HSI is extensively used in various fields like environmental studies [45], military applications [25] and medical fields [36]. The number of bands in HSI varies from 10 to around 400. The main problem with hyperspectral images is the presence of redundant and correlated information, leading to Hughes phenomena [21]. The rich spectral information increases the computing time in processing HSI. Hence, various dimensionality reduction techniques are widely researched [11].

To extract useful information from HSI, different methods like classification, clustering, and unmixing are used [5]. Clustering is a very beneficial method in HSI analysis when the ground truth image is not available. According to Zhang et al. [51], HSI clustering algorithms [32] can be categorized into four groups. Centre-based

DOI: 10.1201/9781003283294-8

method is one of the most widely used clustering algorithms. In this type of clustering, the data points are grouped based on their distance from the cluster centers. K-means [32], is one of the most popular clustering algorithms based on Euclidean distance. It is a hard clustering algorithm, which is sensitive to initial cluster centers and membership values [28]. Fuzzy C means (FCM) [7], is a soft clustering algorithm and produces better results than K-means, at the cost of increased iterations. In both the algorithms, the main disadvantage is that the number of clusters should be mentioned beforehand. In HSI, knowing the number of segments may always not be possible.

Determining the number of clusters in HSI automatically is a challenging task. Recently, researchers have started exploring methods to automatically detect cluster numbers in various problems like image segmentation, data segmentation and others. Very few works have been done on determining the number of clusters in HSI.

Clustering is considered to be a type of nondeterministic hard optimization problem or *NP Hard* problem [4]. Metaheuristics are found to be useful for solving *NP Hard* problems in an efficient manner. Metaheuristics take reasonable time and produce near optimal solutions. Hence, many metaheuristic algorithms have been introduced in the literature for solving clustering problems. Metaheuristic algorithms are stochastic in nature and easy to implement [14]. They are mostly inspired by natural phenomena like swarming of birds [42], ant's colony, food finding behavior [12] and others. Genetic Algorithm (GA) [20], Particle Swarm Optimization (PSO) [42], Ant Colony Optimization [12] and Differential Evolution (DE) [44] are few well-known metaheuristic algorithms.

In recent years, quantum computing has drawn the attention of a lot of researchers. It was originally conceptualized by Sir Richard Feynman [16]. The various quantum phenomena like superposition, entanglement, and interference can enhance the computing capability of an algorithm exponentially [34]. Researchers have explored these ideas to embed the basic principles of quantum computing with metaheuristic algorithms [40]. Hence, a new category of metaheuristics was developed from these, which are called *quantum-inspired metaheuristics*.

The main motivation of this work is to develop a fast and robust automatic clustering algorithm for HSI. Elephant Herd Optimization (EHO) [47] is a comparatively new metaheuristic algorithm, based on the clan formation habit of elephants. The simplicity of the algorithm and its good exploration abilities have inspired in developing the *qubit* and *qutrit* versions of the algorithm called the Qubit Elephant Herd Optimization (QubEHO) and the Qutrit Elephant Herd Optimization (QutEHO), respectively. The exploitation capability of a metaheuristic means that it is capable of refining the result space. The EHO algorithm [47] lacks in this but the quantum versions can easily achieve this. The parallel computing capability of a *qubit* or *qutrit* enhances the exploitation property of the EHO algorithm [47]. Moreover, the QubEHO and QutEHO are found to exhibit higher convergence speeds.

In this work, the Band Selection Convolutional Neural Network (BS-NET-Conv) [8] is used in the pre-processing stage to reduce the number of bands in the HSI. The Xie-Beni Index (XB-Index) [49] is used as the fitness function to detect the optimal number of clusters. The Fuzzy C Means [7] algorithm is used to determine the clusters.

The main contributions of this work are as follows:

- Two algorithms viz., the *qubit* and *qutrit* versions of Elephant Herd Optimization are devised for optimal number of cluster detection in HSI.
- An algorithm for *qubit* based quantum rotation gate implementation for bringing diversity in the population.
- An algorithm for *qutrit* based quantum rotation gate implementation for bringing diversity in the population.

The chapter is organized as follows: Section 8.2 contains a brief literature survey of the used methods. The important background concepts are discussed in Section 8.3. Section 8.4 contains the details of the proposed methodology. The experimental results and their analysis are provided in Section 8.5. A brief conclusion of the proposed method has been drawn in Section 8.6.

8.2 LITERATURE SURVEY

HSI is a 3-dimensional data cube of spectral and spatial information. They are capable of providing detailed information of the captured area. In [53], research on HSI is subdivided into two main categories, viz., supervised methods (classification of HSI [10]) and unsupervised methods (clustering-based algorithms [48]). Dimensionality reduction is also an important area in HSI pre-processing and lots of work are done in this area. Principal Component Analysis [38], Local Discriminant Analysis [30], Mutual Information [15] and Band Selection Convolutional Neural Network (BS-NET-Conv) [8] are few widely used methods in this direction.

Clustering-based algorithms for HSI segmentation have caught the attention of a lot of researchers lately. According to Wang et al. [47], existing clustering algorithms can be classified into four groups, viz., Centroid method [32], Density method [39], Biological methods [54], and Graph methods [52]. *K* means [32] and Fuzzy *C* means [7] are examples of centroid methods. They are easy to implement but are not always capable of providing robust results.

To determine the optimal number of clusters, CVI is used. A cluster validity index (CVI) defines a relation between intracluster cohesion (within-group scatter) and intercluster separation (between-group scatter) to estimate the quality of a clustering solution [24]. Xie-Beni Index [49], Dunn index [13], Calinski - Harabasz Index [9], Gamma index [6], I-Index [33] are few widely used indices. In [24], the authors have provided an extensive study on CVI for automatic data clustering.

Finding the optimal number of clusters is a type of *NP Hard* problem. To solve these problems, metaheuristic algorithms are found to be highly beneficial. Metaheuristic algorithms are developed based on some natural phenomena. They can be widely classified into four categories [14], viz., Evolutionary Algorithms [20], Physics-based Algorithms [26], Human-based Algorithms [37], and Swarm Algorithms [42].

Swarm algorithms are a popular category of metaheuristic algorithms. They are mostly inspired by the social behavior of animals or insects. The ability of a swarm to reach an optimal solution collectively and using individual intelligence makes swarm a powerful tool for optimization. Particle Swarm Optimization [42], Ant Colony Optimization [12], Cuckoo Search [50], Harris hawk Optimization [19], Border Collie Optimization [14] and Elephant Herd Optimization [47] are few well-known swarm intelligent algorithms. EHO [47], due to its easy implementation and good performance, has drawn the attention of a lot of researchers. The algorithm has good exploration capability but the exploitation of search space is less efficient. This also leads to slower convergence. Hence, a lot of enhanced and hybrid versions of the EHO algorithm have been researched [23] to overcome these disadvantages. In [23], three different enhanced versions of EHO algorithm were proposed to overcome these deficiencies. Li et al. [29] presented a detailed study on the different variants of the EHO algorithm published so far, along with their different features.

Researchers working on metaheuristic algorithms, have been captivated by the idea of designing algorithms with quantum advantage. These are also known as quantum-inspired algorithms, as they are inspired by the principles of quantum computing but the simulations are done on classical computers. Narayanan and Moore were the first to conceptualize a quantum-inspired evolutionary algorithm for solving the Travelling Salesman problem [35]. In [18], a Quantum-Inspired Evolutionary Algorithm was proposed with a better population diversity and concept of look-up table for the application of rotation gates. Recently, a lot of researchers has developed quantum-inspired versions of metaheuristic algorithms like the Improved Bloch Quantum Artificial Bee Colony algorithm, which was proposed in [22]. It involves a complicated Bloch sphere representation, which makes it a little complex. Very few works have been done on multilevel quantum systems. A *qutrit*-based Genetic Algorithm was proposed in [46]. However, hardly any work has been done yet on developing multilevel quantum-based EHO algorithm.

8.3 BACKGROUND CONCEPTS

A few important background concepts used for developing the proposed algorithm are presented in this section.

8.3.1 ELEPHANT HERDING OPTIMIZATION

Elephant Herding Optimization [47] is a new nature-inspired metaheuristic algorithm based on the herding behavior of elephants. The behavioral factors considered to develop the mathematical model were:

- All elephants live in clans under the leadership of a matriarchal elephant.
- The number of elephants in a clan is fixed.
- Male elephants leave their clan and live alone after a certain time.

8.3.1.1 Clan Updation

For developing the EHO algorithm [47], initially few clans are considered. The matriarch ($Ebest$) is the fittest elephant of the clan. Her position is updated using the following equation:

$$Ebest_{cl_i} = \beta \times E_{c,cl_i} \tag{8.1}$$

In Eqn. (8.1), cl_i is the ith clan, $Ebest_{cl_i}$ represents the updated position of the matriarch in the cl_i^{th} clan, β is a factor that dictates the influence of the E_{c,cl_i} on new $Ebest_{new,cl_i,j}$. The value of β ranges between $[0,1]$. E_{c,cl_i} is the center of clan cl_i, calculated with the help of the following equation.

$$E_{c,cl_i,d} = \frac{1}{n_{cl_i}} \times \sum_{j=1}^{n_{ci}} E_{cl_i,j,d} \tag{8.2}$$

In Eqn. (8.2), d is the dth dimension and D is the total dimension. n_{cl_i} is the number of elephants in the cl_i^{th} clan. Eqn. (8.2) is used to calculate the center of a clan.

All the other elephants (j) next positions are calculated using the following equation:

$$E_{new,cl_i,j} = E_{cl_i,j} + \alpha \times (Ebest_{cl_i} - E_{cl_i,j}) \times r \tag{8.3}$$

In Eqn. (8.3), α is the influence that the matriarch has on the other elephants of the clan. Its value ranges from $[0,1]$. r is a random number.

8.3.1.2 Separation Operator

The male elephants on reaching puberty usually leave the clan and live on their own. Usually, the most unfit individual is considered as the male elephant leaving the clan. Its position is updated in the following manner:

$$E_{worst,cl_i} = E_{min} + (E_{max} - E_{min} + 1) \times \text{rand} \tag{8.4}$$

where, E_{max} and E_{min} are the upper and lower bounds of the elephant individuals. E_{worst,cl_i} is the worst elephant individual in the ith clan. *rand* is a random number generated between $[0,1]$.

8.3.1.3 Steps of EHO

The EHO algorithm [47] works in the following manner:

- Initialize an elephant population.
- Run the following steps for *Ge* number of generations.
    - Find the fitness of the elephants and sort them.
    - Implement clan updation using Eqns. (8.1),(8.2) and (8.3)
    - Implement separation operator

8.3.2 BASIC CONCEPTS OF QUANTUM COMPUTING

A *qubit* is the smallest unit of quantum computation [34]. A physical system that can be present in at least two orthogonal states can be used to represent a *qubit*. It can be a spin particle in a magnetic field or a photon in polarized state [43]. The orthogonal states are represented as $|0\rangle$ and $|1\rangle$. This notation is called the Dirac's Bra and Ket notation. Applying the laws of quantum mechanics, the orthogonal states can be represented in their linear combination.

$$|\psi\rangle = q_0|0\rangle + q_1|1\rangle \tag{8.5}$$

In Eqn. (8.5), q_0 is the probability of occurrence of $|0\rangle$ and q_1 is the probability of occurrence of $|1\rangle$. The probability constraints, q_0 and q_1 are complex numbers, which satisfy the normalization condition as given below.

$$|q_0|^2 + |q_1|^2 = 1 \qquad (8.6)$$

This property of *qubits* is also called quantum superposition in which they are capable of existing in both basis states simultaneously. Hence, if an *n qubit* system is considered, it will be able to exist in 2^n states simultaneously. This can be explained with the help of the following equation:

$$|\psi\rangle = \sum_{i=0}^{2^n-1} q_i|i\rangle \qquad (8.7)$$

Quantum computation like classical computation is not restricted to two states. Quantum states can have D-dimensional states. This generalized state of quantum bits is called *qudit* [3]. Hence, a *qudit* can exist in the following state.

$$|\psi\rangle = \sum_{i=0}^{D^n-1} q_i|i\rangle \qquad (8.8)$$

A *qutrit* is the three-valued quantum state. It has three basis states, viz., $|0\rangle$, $|1\rangle$ and $|2\rangle$. The superposition of a *qutrit* state can be expressed as follows:

$$|\psi\rangle = q_0|0\rangle + q_1|1\rangle + q_2|2\rangle \qquad (8.9)$$

The normalization of a *qutrit* state is expressed as follows:

$$|q_0|^2 + |q_1|^2 + |q_2|^2 = 1 \qquad (8.10)$$

8.3.3 FUZZY C MEANS CLUSTERING ALGORITHM

Fuzzy C Means [7] is one of the most frequently used clustering algorithms. It is a soft clustering method in which a pixel is given membership values based on the different cluster centers [5]. The main aim of the *FCM* algorithm [7] is to find optimal clusters, by minimizing the squared function given below [17].

$$M = \sum_{i=1}^{k}\sum_{j=1}^{d} U_{ij}^m d_{ij}^2 = \sum_{i=1}^{k}\sum_{j=1}^{d} U_{ij}^m \left\| P_j - V_i \right\|^2 \qquad (8.11)$$

In Eqn. (8.11), U_{ij} represents the membership values of a pixel to k cluster centers. The value of U_{ij} varies between 0 and 1. d represents the total number of pixels of the image that is being clustered and m is the weighting exponent which ranges between $[1, \infty]$. V_i represents the k cluster centers. $d_{i,j}$ is the Euclidean distance between the ith cluster center and the jth data point and P_j is the jth data point. V_i is calculated using the following equation:

$$V_i = \frac{\sum_{j=1}^{d} U_{ij}^m P_j}{\sum_{j=1}^{d} U_{ij}^m} \qquad (8.12)$$

The membership values are calculated as given below.

$$U_{ij} = \frac{1}{\sum_{i=1}^{k} \left(\frac{d_i}{d_{ji}}\right)^{2/(m-1)}} = \left(\sum_{i=1}^{k} \left(\frac{\|P_j - V_i\|}{\|P_j - V_i\|}\right)^{2/(m-1)}\right)^{-1} \tag{8.13}$$

The process is run for a number of iterations until a minimum value for Eqn. (8.11) is obtained.

8.3.4 XIE-BENI INDEX

Xie-Beni Index is used for determining the optimum number of clusters [49]. It is a quotient of mean quadratic error and minimal squared distance between the points in the cluster. The XB-Index is represented as follows:

$$XB = \frac{M_2}{d_{\min}} \tag{8.14}$$

where, the compactness of a cluster is represented as follows:

$$M_m = \frac{1}{d} \sum_{i=1}^{k} \sum_{j=1}^{d} u_{ij}^m d^2 \left(P_j, k_i\right) \tag{8.15}$$

and the minimum distance between two cluster centers is given by the following equation:

$$d_{\min} = \min_{i,j} \left[d^2 \left(k_i, k_j\right)\right] \tag{8.16}$$

XB-Index [49] is a minimization function and the minimum value of Eqn. (8.14), represents the optimal solution.

8.4 PROPOSED METHODOLOGY

The proposed method can be divided into two parts, viz., the hyperspectral image pre-processing and the quantum versions of the Elephant Herd Optimization [47] algorithm based determination of the optimal number of clusters.

8.4.1 HSI PREPROCESSING

The Band Selection Convolution Neural Network [8] is implemented in the pre-processing stage, to reduce the number of bands in HSI. This algorithm chooses the best informative bands, which are then used in the automatic clustering stage. The network used, consists of a band attention module and a band re-weighing module. This is followed by a reconstruction network. The attention module extracts information from the HSI cube using 2-D convolution layers. To ensure that the learned weights are non-negative, the Sigmoid function is used. Finally, a global pooling layer and a fully connected layer are implemented. The best bands are selected based on the average of weights of all the bands.

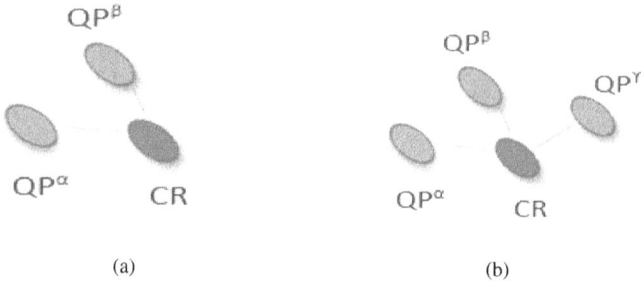

Figure 8.1: Qubit and Qutrit representation in search space.

8.4.2 QUBIT AND QUTRIT BASED ELEPHANT HERD OPTIMIZATION

The proposed Qubit and Qutrit Elephant Herd Optimization are depicted with the help of Algorithm 1. Initially, a quantum population consisting QP is taken consisting of $|QP^\alpha\rangle, |QP^\beta\rangle$ for *qubit* or $|QP^\alpha\rangle, |QP^\beta\rangle, |QP^\gamma\rangle$ for *qutrit*.

Result: *Optimal Number of Clusters(k)*
Initialize: Number of Generation - *Ge*,
        Size of Quantum Population - *n*,
        Length of each particle -*D*,
        No of clans - *cl*,
        Qubit/Qutrit population initialization and corresponding
        classical representation - *QP* and *CR* (using Algorithm 2 for
        qubit or using Algorithm 3 for *qutrit*)
Applying Rotation Gate on *QP* using Algorithm 4 for *qubit* or using Algorithm 5 for *qutrit*
Allotting the elephants to *cl* different clans randomly
for *t in 1,2,…,Ge* **do**
    z=Random number of zeros; Randomly insert *z* zeros in *CR*; Calculate the *z* number of clusters using
    FCM algorithm [7]; Calculate Fitness of Population using $XB - Index$ - Fit_t; **for** *i in 1,2,…,cl* **do**
        Update *QP* value for Matriarch of the clan *i* ((QP^α or QP^β) for *qubit* or (QP^α or QP^β or QP^γ) for
        qutrit, for which *CR* is true) using Eqn. (8.1);
        Update *QP* value for other elephants of the clan *i* ((QP^α or QP^β) for *qubit* or (QP^α or QP^β or
        QP^γ) for *qutrit*, for which *CR* is true) using Eqn. (8.3);
        Update *QP* value for most unfit elephant of the clan *i* ((QP^α or QP^β) for *qubit* or (QP^α or QP^β or
        QP^γ) for *qutrit*, for which *CR* is true) using Eqn. (8.4);
    end
    if $Fit_t < Fit_{t-1}$ **then**
        k=Optimal cluster number obtained in t^{th} generation;
    Update the quantum states to maintain superposition principle;
end

Algorithm 1: QubEHO/QutEHO

$$|QP\rangle = \frac{1}{\sqrt{2}}|0\rangle + \frac{1}{\sqrt{2}}|1\rangle \qquad (8.17)$$

$$|QP\rangle = \frac{1}{\sqrt{3}}|0\rangle + \frac{1}{\sqrt{3}}|1\rangle + \frac{1}{\sqrt{3}}|2\rangle \qquad (8.18)$$

Their corresponding basis state predictions (*CR*) are done using Algorithm 2 for *qubit* or Algorithm 3 for *qutrit*. Algorithm 4 for *qubit* or Algorithm 5 for *qutrit* are

Result: Basis states -$|0\rangle,|1\rangle$
Initialize: QP using Eqn. (8.17)
for i *in 1,2,...,n* **do**
    for j *in 1,2,...,D* **do**
        r=random number between [0,1];
        if $r < \alpha^2$ **then**
         | $CR = 0$;
        else
         | $CR = 1$;
        end
    end
end

Algorithm 2: Classical Representation of Quantum States

Result: Basis states -$|0\rangle,|1\rangle,|2\rangle$
Initialize: QP using Eqn. (8.18)
for i *in 1,2,...,n* **do**
    for j *in 1,2,...,D* **do**
        r=random number between [0,1];
        if $r < \alpha^2$ **then**
         | $CR = 0$;
        else if $r1 < \beta^2 + \alpha^2$ **then**
         | $CR = 1$;
        else
         | $CR = 2$;
        end
    end
end

Algorithm 3: Classical Representation of Quantum States

proposed for rotating the quantum states without the help of look-up-tables. The implementation of rotation gates brings diversity to the population. Each *qubit* or

Result: Rotated states -$|QP^\alpha\rangle,|QP^\beta\rangle$
for i *in 1,2,...,n* **do**
    for j *in 1,2,...,D* **do**
        r=random number between [0,1];
        if $CR_{i,j} == 0$ **then**
         | $QP^\alpha_{i,j} = QP^\alpha_{i,j} * r; QP^\beta_{i,j} = \sqrt{1 - QP^{\alpha 2}_{i,j}}$;
        else if $CR_{i,j} == 1$ **then**
         | $QP^\beta_{i,j} = QP^\beta_{i,j} * r; QP^\alpha_{i,j} = \sqrt{1 - QP^{\beta 2}_{i,j}}$;
    end
end

Algorithm 4: Rotation of Quantum States

qutrit represents the individual elephant. A fixed number of clans are considered and the population is scattered randomly among the clans. Then, a random number of zeros (z) are introduced in CR. The number of zeros represents the number of clusters to be considered. The QP value for which CR is true, is taken as the cluster center. z number of clusters are calculated using FCM algorithm [7]. The *XB-Index* [49] is used as the fitness function to check the optimality of the number of clusters obtained.

Subsequently, the clan updation operation is done. The Matriarch of each clan is identified. The QP value for which CR is true, is updated using Eqn. (8.1). Other elephants of the clan are updated similarly using Eqn. (8.2). The separation operation

Result: Rotated states $-|QP^{\alpha}\rangle, |QP^{\beta}\rangle, |QP^{\gamma}\rangle$

for i in $1,2,\ldots,n$ **do**

    for j in $1,2,\ldots,D$ **do**

        r=random number between $[0,1]$;

        if $CR_{i,j} == 0$ **then**

            $QP^{\alpha}_{i,j} = QP^{\alpha}_{i,j} * r;\ QP^{\beta}_{i,j} = \sqrt{1 - QP^{\alpha 2}_{i,j}};\ QP^{\gamma}_{i,j} = \sqrt{1 - QP^{\alpha 2}_{i,j} + QP^{\beta}_{i,j}};$

        else if $CR_{i,j} == 1$ **then**

            $QP^{\beta}_{i,j} = QP^{\beta}_{i,j} * r;\ QP^{\alpha}_{i,j} = \sqrt{1 - QP^{\beta 2}_{i,j}};\ QP^{\gamma}_{i,j} = \sqrt{1 - QP^{\alpha 2}_{i,j} + QP^{\beta}_{i,j}};$

        else

            $QP^{\gamma}_{i,j} = QP^{\gamma}_{i,j} * r;\ QP^{\alpha}_{i,j} = \sqrt{1 - QP^{\gamma 2}_{i,j}};\ QP^{\beta}_{i,j} = \sqrt{1 - QP^{\alpha 2}_{i,j} + QP^{\gamma}_{i,j}};$

        end

    end

end

Algorithm 5: Rotation of Quantum States

is executed on the least fit elephant using Eqn. (8.4). The whole process is run for *Ge* number of generations. In Figure 8.2, the basic operations viz., formation of the clan, separation of male elephant from the clan and how other elephants follow the Matriarch can be visualized. The use of quantum bits helps to search the solution space faster and more efficiently. This can be visualized in Figure 8.1. For a single *qubit*, we can see that two different solutions points are considered. Simiarly, for a single *qutrit* three different solutions points are taken. Hence, it enhances the speed of the algorithm producing more robust results.

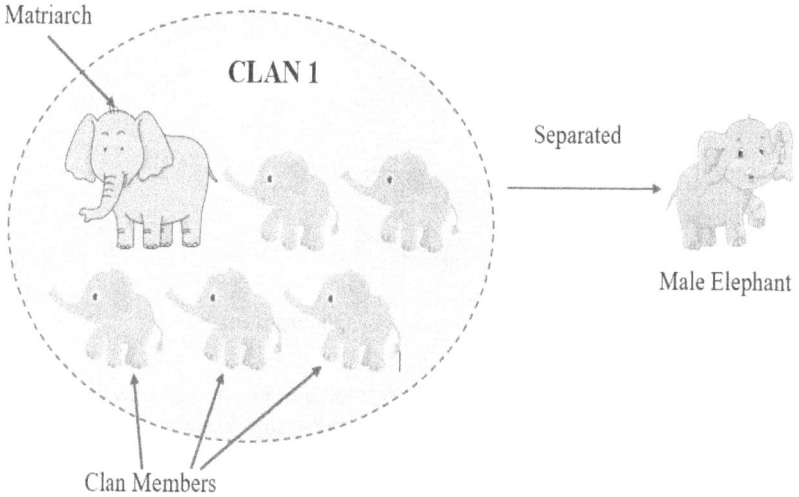

(a) Elephant Herd Optimization

Figure 8.2: Elephant Herd Optimization [47]-Clan formation, Clan members following matriarch, Seperation from clan (Elephant picture from [1]).

8.5   EXPERIMENTAL RESULTS AND ANALYSIS

The Salinas Dataset [2] is used for experimental purpose. A brief description of the dataset is presented in Section 8.5.1. The proposed QubEHO and QutEHO algorithms are compared with classical EHO [47] algorithm. Section 8.5.2 contains a brief description of the fitness function used. The different parameters used to analyze the results are presented in Section 8.5.3.

8.5.1   SALINAS DATASET

The Salinas scene was collected by the AVIRIS sensor over the Salinas Valley California [2]. A total of 224 bands were captured and 20 were discarded due to water absorption. The spatial resolution is 3.7m pixels consisting of 512×217 bands each. The ground truth image of the Salinas dataset [2] consists of 16 classes, which includes vegetables, bare soils, and vineyards.

8.5.2   FITNESS FUNCTION

The *Xie-Beni Index* [49] is used as the fitness function for determining the optimal number of clusters. Eqn. (8.14) is minimized for obtaining the optimal results.

8.5.3   ANALYSIS

The proposed QubEHO and QutEHO algorithms are compared with classical EHO [47] algorithm. To obtain an impartial analysis, all the algorithms need to be evaluated using the same parameters. Hence, all the algorithms were run 50 times for 100 iterations each. The algorithms were executed in MATLAB R2019a, on Intel (R) Core (TM) i7 8700 Processor in Windows 10 environment.

The mean, standard deviation, and best convergence time of all the three algorithms are presented in Table 8.1. It is observed that the proposed QutEHO algorithm arrives at optimal results in negligible time when compared to the other two algorithms. The classical EHO [47] algorithm takes the highest time to converge.

In Table 8.2, the optimal cluster numbers (CL) obtained for EHO [47], QubEHO, and QutEHO are reported. Their corresponding fitness values (FV) are also presented in Table 8.2 where . The Salinas dataset [2], has 16 classes. The most optimal results

Table 8.1

Mean, Standard Deviation (Std) and the Best Reported Time for EHO [47], QubEHO, and QutEHO

Parameters	EHO [47]	QubEHO	QutEHO
Mean	0.2032	0.0725	0.0389
Std	0.1394	0.0124	0.0219
Time	181.9412	24.3839	**3.3035**

Table 8.2

Some Cluster Numbers and Best Fitness Values Obtained for EHO [47], QubEHO and QutEHO.

Methods	EHO [47]		QubEHO		QutEHO	
Sr No	CL	FV	CL	FV	CL	FV
1.	4	0.0685	6	0.0733	**10**	**0.0129**
2.	4	0.0828	5	0.0693	9	0.0143
3.	4	0.0835	5	0.0787	9	0.0144
4.	5	0.0835	5	0.0660	7	0.0182
5.	4	0.0860	5	0.0614	7	0.0182

Table 8.3

Normalized F-Score [31] for EHO [47], QubEHO and QutEHO

Process	EHO [47]	QubEHO	QutEHO
	0.1998	0.0038	**0.0016**

are produced by the QutEHO algorithm, followed by the QubEHO algorithm. The classical EHO [47] has comparatively lesser number of classes. In a real-life scenario, where the ground truth cannot be obtained and the number of classes cannot be estimated, the proposed algorithms, specially the *qutrit* version can be both beneficial and time efficient.

To judge the quality of the segmented images, the normalized F score is used [31]. The F score is evaluated with the help of the following equation.

$$F(SI) = \frac{1}{1000(v \times h)} \sqrt{r} \sum_{i=1}^{r} \frac{e_i^2}{\sqrt{A_i}} \qquad (8.19)$$

Here, SI is the final segmented image. The dimension of the image is $v \times h$. The number of regions is designated by r. A and e stand for the area and the average color error of the ith region, respectively. The results are normalized for easy representation. The results of all the three methods, viz., EHO [47], QubEHO, and QutEHO are presented in Table 8.3. The proposed algorithm produces better results compared to the other two algorithms. Few segmented images along with the ground truth image and the pre-processed images are presented in Figure 8.3. In Figure 8.3, the segmented images using EHO [47] has four clusters, QubEHO has five clusters and QutEHO has nine clusters, respectively.

Another statistical test called the Kruskal-Wallis test [27] is applied to check the null hypothesis with 1% significance level. The p-value obtained is less than 0.001 indicating that the results are highly significant. Hence, the null hypothesis that all the results from all the three methods belong to the same distribution, stands rejected.

(a) Ground Truth Image   (b) Pe-processed Image   (c) EHO [47]

(d) QubEHO   (e) QutEHO

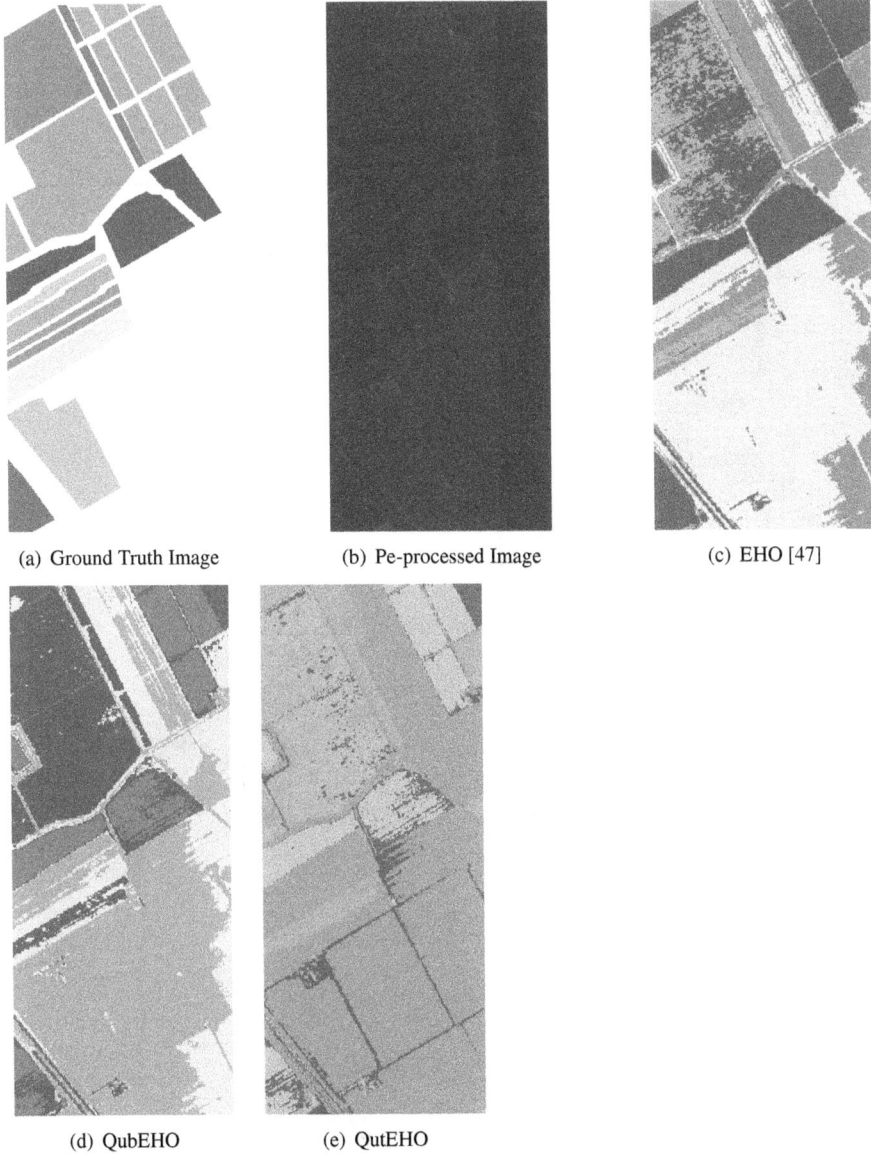

Figure 8.3: (a) Ground Truth Image of Salinas Dataset [2], (b) Pre-processed Image using BSCNN [8], (c)-(e) Clustered Images using EHO [47], QubEHO, and QutEHO.

The results are presented in Table 8.4. The box-plot of the test is given in Figure 8.4.

The convergence curve of all the participating methods is presented in Figure 8.5. From the convergence curve, it can be visually observed that the proposed algorithm converges faster and with more optimal values compared to EHO [47] and QubEHO.

Table 8.4

Kruskal-Wallis Test [27]

Test	p-value	Significance
Kruskal-Wallis Test	2.8166e-25	Highly Significant

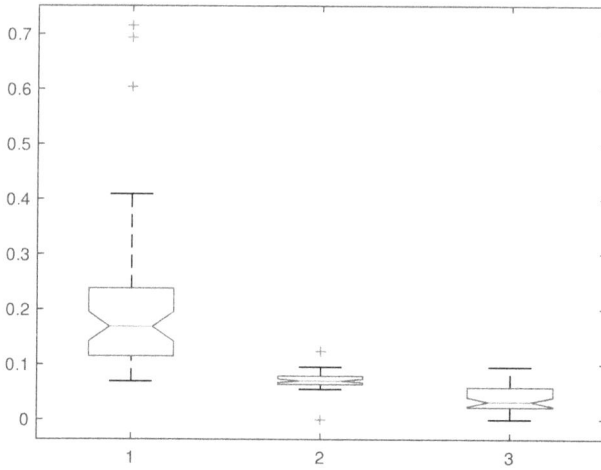

(a) Box-Plot of Kruskal-Wallis Test [27]

Figure 8.4: Box-Plot of Kruskal-Wallis Test [27] for EHO [47], QubEHO, and

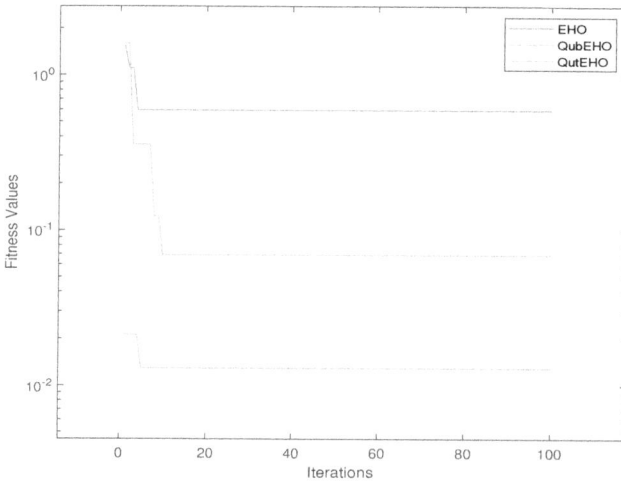

(a) Convergence Curve

Figure 8.5: Convergence curve for EHO [47], QubEHO, and QutEHO.

8.6 CONLUSION

In this chapter, *qubit* and *qutrit*-based Elephant Herd Optimization algorithms are proposed for automatic clustering of hyperspectral images. The modified rotation gate operation enhances the diversity of the population. The exploration and exploitation capabilities of the classical EHO algorithm is enhanced, with a faster convergence capability. As automatic cluster detection is a tedious task in HSI processing, these algorithms can be highly beneficial in real-life scenarios. The results indicate that the proposed QubEHO, and QutEHO produce optimal results compared to the classical version of EHO. The proposed algorithms produces better clustering outcome when their comparative *F* scores are considered. The statistical tests also establish the efficiency of the proposed algorithms. As a future direction, the *qudit* version of the EHO algorithm can be developed.

REFERENCES

1. Elephant clipart. Free download transparent .PNG | Creazilla, Jul 2021. [Online; accessed 17. Jul. 2021].
2. Hyperspectral Remote Sensing Scenes - Grupo de Inteligencia Computacional (GIC), Jul 2021. [Online; accessed 17. Jul. 2021].
3. Qudits | Cirq | Google Quantum AI, Jul 2021. [Online; accessed 16. Jul. 2021].
4. Laith Abualigah, Amir H. Gandomi, Mohamed Abd Elaziz, Husam Al Hamad, Mahmoud Omari, Mohammad Alshinwan, and Ahmad M. Khasawneh. Advances in meta-heuristic optimization algorithms in big data text clustering. *Electronics*, 10(2), 2021.
5. P. Azimpour, R. Shad, M. Ghaemi, and H. Etemadfard. Hyperspectral image clustering with albedo recovery fuzzy c-means. *International Journal of Remote Sensing*, 41(16):6117–6134, 2020.
6. Frank B. Baker and Lawrence J. Hubert. Measuring the power of hierarchical cluster analysis. *Journal of the American Statistical Association*, 70(349):31–38, 1975.
7. J. C. Bezdek, R. Ehrlich, and W. Full. Fcm: The fuzzy c-means clustering algorithm. *Computers & Geosciences*, 10(2):191 – 203, 1984.
8. Y. Cai, X. Liu, and Z. Cai. Bs-nets: An end-to-end framework for band selection of hyperspectral image. *IEEE Transactions on Geoscience and Remote Sensing*, 58(3):1969–1984, 2020.
9. T. Caliński and H. JA. A dendrite method for cluster analysis. *Communications in Statistics - Theory and Methods*, 3:1–27, 01 1974.
10. Mulin Chen, Qi Wang, and Xuelong Li. Discriminant analysis with graph learning for hyperspectral image classification. *Remote Sensing*, 10(6), 2018.
11. Samiran Das, Shubhobrata Bhattacharya, Aurobinda Routray, and Alok Kani Deb. Band selection of hyperspectral image by sparse manifold clustering. *IET Image Processing*, 13(10):1625–1635, 2019.
12. M. Dorigo, V. Maniezzo, and A. Colorni. Ant system: optimization by a colony of cooperating agents. *IEEE Transactions on Systems, Man, and Cybernetics, Part B (Cybernetics)*, 26(1):29–41, 1996.
13. J. C. Dunn. A fuzzy relative of the isodata process and its use in detecting compact well-separated clusters. *Journal of Cybernetics*, 3(3):32–57, 1973.
14. Tulika Dutta, Siddhartha Bhattacharyya, Sandip Dey, and Jan Platos. Border collie optimization. *IEEE Access*, 8:109177–109197, 2020.

15. A. Elmaizi, H. Nhaila, E. Sarhrouni, A. Hammouch, and C. Nacir. A novel information gain based approach for classification and dimensionality reduction of hyperspectral images. *Procedia Computer Science*, 148:126 – 134, 2019.

16. R.P. Feynman. Simulating physics with computers. *International Journal of Theoretical Physics*, 21(6):467–488, 1982.

17. Saman Ghaffarian and Salar Ghaffarian. Automatic histogram-based fuzzy c-means clustering for remote sensing imagery. *ISPRS Journal of Photogrammetry and Remote Sensing*, 97:46–57, 2014.

18. K-H Han and J-H Kim. Quantum-inspired evolutionary algorithm for a class of combinatorial optimization. *IEEE Transactions on Evolutionary Computation*, 6(6):580–593, 2002.

19. Ali Asghar Heidari, Seyedali Mirjalili, Hossam Faris, Ibrahim Aljarah, Majdi Mafarja, and Huiling Chen. Harris hawks optimization: Algorithm and applications. *Future Generation Computer Systems*, 97:849–872, 2019.

20. J.H. Holland. *Adaptation in natural and artificial systems : an introductory analysis with application to biology. Control and artificial intelligence.* University of Michigan Press, 1975.

21. G. Hughes. On the mean accuracy of statistical pattern recognizers. *IEEE Transactions on Information Theory*, 14(1):55–63, 1968.

22. F. Huo, X. Sun, and W. Ren. Multilevel image threshold segmentation using an improved bloch quantum artificial bee colony algorithm. *Multimedia Tools and Applications*, 79:24472471, 2020.

23. Alaa A. K. Ismaeel, Islam A. Elshaarawy, Essam H. Houssein, Fatma Helmy Ismail, and Aboul Ella Hassanien. Enhanced elephant herding optimization for global optimization. *IEEE Access*, 7:34738–34752, 2019.

24. Adán José-García and Wilfrido Gómez-Flores. A survey of cluster validity indices for automatic data clustering using differential evolution. In *Proceedings of the Genetic and Evolutionary Computation Conference*, GECCO '21, page 314322, New York, NY, USA, 2021. Association for Computing Machinery.

25. Chen Ke. Military object detection using multiple information extracted from hyperspectral imagery. In *2017 International Conference on Progress in Informatics and Computing (PIC)*, pages 124–128, 2017.

26. S. Kirkpatrick, C. D. Gelatt, and M. P. Vecchi. Optimization by simulated annealing. *Science*, 220(4598):671–680, 1983.

27. W. H. Kruskal and W. A. Wallis. Use of ranks in one-criterion variance analysis. *Journal of the American Statistical Association*, 47(260):583–621, 1952.

28. Tao Lei, Xiaohong Jia, Yanning Zhang, Shigang Liu, Hongying Meng, and Asoke K. Nandi. Superpixel-based fast fuzzy c-means clustering for color image segmentation. *IEEE Transactions on Fuzzy Systems*, 27(9):1753–1766, 2019.

29. Juan Li, Hong Lei, Amir H. Alavi, and Gai-Ge Wang. Elephant herding optimization: Variants, hybrids, and applications. *Mathematics*, 8(9), 2020.

30. W. Liao, A. Pizurica, P. Scheunders, W. Philips, and Y. Pi. Semisupervised local discriminant analysis for feature extraction in hyperspectral images. *IEEE Transactions on Geoscience and Remote Sensing*, 51(1):184–198, 2013.

31. Jianqing Liu and Yee-Hong Yang. Multiresolution color image segmentation. *IEEE Transactions on Pattern Analysis and Machine Intelligence*, 16(7):689–700, 1994.

32. J. Macqueen. Some methods for classification and analysis of multivariate observations. In *Proceedings of the Fifth Berkeley Symposium on Mathematical Statistics and Probability, Volume 1: Statistics*, pages 281–297, 1967.

33. U. Maulik and S. Bandyopadhyay. Performance evaluation of some clustering algorithms and validity indices. *IEEE Transactions on Pattern Analysis and Machine Intelligence*, 24:1650–1654, 01 2002.
34. D. McMahon. *Quantum Computing Explained*. John Wiley & Sons, Inc., Hoboken,New Jersey, 2008.
35. A. Narayanan and M. Moore. Quantum-inspired genetic algorithms. In *Proceedings of IEEE International Conference on Evolutionary Computation*, pages 61–66, 1996.
36. Robert Pike, Guolan Lu, Dongsheng Wang, Zhuo Georgia Chen, and Baowei Fei. A minimum spanning forest-based method for noninvasive cancer detection with hyperspectral imaging. *IEEE Transactions on Biomedical Engineering*, 63(3):653–663, 2016.
37. T. Ray and K.M. Liew. Society and civilization: An optimization algorithm based on the simulation of social behavior. *IEEE Transactions on Evolutionary Computation*, 7(4):386–396, 2003.
38. Craig Rodarmel and Jie Shan. Principal component analysis for hyperspectral image classification. *Surv Land inf Syst*, 62, 01 2002.
39. Alex Rodriguez and Alessandro Laio. Clustering by fast search and find of density peaks. *Science*, 344(6191):1492–1496, 2014.
40. Maria Schuld, Ilya Sinayskiy, and Francesco Petruccione. An introduction to quantum machine learning. *Contemporary Physics*, 56(2):172–185, 2015.
41. G. Shaw and Hsiao hua K. Burke. Spectral imaging for remote sensing. In *Lincoln Laboratory Journal*, number 1, pages 3–28, 2003.
42. Y. Shi and R. Eberhart. A modified particle swarm optimizer. In *1998 IEEE International Conference on Evolutionary Computation Proceedings. IEEE World Congress on Computational Intelligence (Cat. No.98TH8360)*, pages 69–73, May 1998.
43. M. Steffen, D. P. DiVincenzo, J. M. Chow, T. N. Theis, and M. B. Ketchen. Quantum computing: An ibm perspective. *IBM Journal of Research and Development*, 55(5):13:1–13:11, 2011.
44. Rainer Storn and Kenneth Price. Differential evolution – a simple and efficient heuristic for global optimization over continuous spaces. *J. of Global Optimization*, 11(4):341–359, 1997.
45. Ming Yang Teng, Ruby Mehrubeoglu, Scott A. King, Kirk Cammarata, and James Simons. Investigation of epifauna coverage on seagrass blades using spatial and spectral analysis of hyperspectral images. In *2013 5th Workshop on Hyperspectral Image and Signal Processing: Evolution in Remote Sensing (WHISPERS)*, pages 1–4, 2013.
46. V. Tkachuk. Quantum genetic algorithm based on qutrits and its application. *Mathematical Problems in Engineering*, 2018(8614073), 2018.
47. Gai-Ge Wang, Suash Deb, and Leandro dos S. Coelho. Elephant herding optimization. In *2015 3rd International Symposium on Computational and Business Intelligence (ISCBI)*, pages 1–5, 2015.
48. Huan Xie, Ang Zhao, Shengyu Huang, Jie Han, Sicong Liu, Xiong Xu, Xin Luo, Haiyan Pan, Qian Du, and Xiaohua Tong. Unsupervised hyperspectral remote sensing image clustering based on adaptive density. *IEEE Geoscience and Remote Sensing Letters*, 15(4):632–636, 2018.
49. X. L. Xie and G. Beni. A validity measure for fuzzy clustering. *IEEE Transactions on Pattern Analysis and Machine Intelligence*, 13(8):841–847, Aug 1991.
50. Xin-She Yang and Suash Deb. Cuckoo search via lvy flights. In *2009 World Congress on Nature Biologically Inspired Computing (NaBIC)*, pages 210–214, 2009.

51. Hongyan Zhang, Han Zhai, Liangpei Zhang, and Pingxiang Li. Spectralspatial sparse subspace clustering for hyperspectral remote sensing images. *IEEE Transactions on Geoscience and Remote Sensing*, 54(6):3672–3684, 2016.

52. Yang Zhao, Yuan Yuan, Feiping Nie, and Qi Wang. Spectral clustering based on iterative optimization for large-scale and high-dimensional data. *Neurocomputing*, 318:227–235, 2018.

53. Yang Zhao, Yuan Yuan, and Qi Wang. Fast spectral clustering for unsupervised hyperspectral image classification. *Remote Sensing*, 11(4), 2019.

54. Yanfei Zhong, Liangpei Zhang, and Wei Gong. Unsupervised remote sensing image classification using an artificial immune network. *International Journal of Remote Sensing*, 32(19):5461–5483, 2011.

9 Toward Quantum-Inspired SSA for Solving Multiobjective Optimization Problems

9.1 INTRODUCTION

Many real-world optimization problems are multiobjective optimization problems. It must find a uniformly distributed set of optimal solutions along the Pareto-optimal front in the search space. Multiobjective optimization problem (MOP) is an empirical area of research. It has been applied in many sciences and engineering fields, including mathematical optimization and the controller placement problem, where ideal choice requires to be made in the existence of trade-offs between two or more conflicting objectives. In general, a multiobjective optimization problem can be a minimization or maximization problem $f(x)$ subject to $x \varepsilon X$, where $f(x)$ is a real-valued scalar function often known as objective or cost function and X is the feasible set or subset of the search space. Without the loss of assumptions, it can be mathematically presented as below:

$$\begin{cases} \text{minimize } f(x) = (f_1(x), f_1(x), \ldots f_n(x)) \\ \quad where \ x \ \varepsilon \ X \\ \quad\quad n \geq 2 \\ \text{set X includes all the feasible solutions} \end{cases}$$

where $x = (x_1, x_1, \ldots x_d)$ is a d-dimensional vector in the search space R^d, $f(x) = (f_1(x), f_1(x), \ldots f_n(x))$ is an n-dimensional vector in the search space R^n. After all, the objectives in MOP are commonly counter to one and all. The Pareto dominion association is often applied to differentiate distinct solutions. According to the Pareto dominion association, for any two given solutions x and x', it can be said that x dominates another solution x',

if $\forall i \varepsilon \{1,2,3,\ldots n\}$, $f_i(x) \leq f_i(x')$ and
$\exists i \varepsilon \{1,2,3,\ldots n\}$, $f_i(x) < f_i(x')$

In case solution x dominates any other solutions, x, x could be said as a non-dominated solution. The set of all other than dominated solutions in the search space

DOI: 10.1201/9781003283294-9

is called estimated Pareto-optimal (PE) [1]. The set of vectors in the search space that correspond to the PE is mentioned as the actual Pareto-optimal (PA) [2][3]. The essence of solving a multiobjective optimization problem is identifying a set of well-distributed optimal solutions along the Pareto-optimal front in the search space.

Today, a metaheuristic algorithm is a popular approach to resolve compound optimization problems in the field of optimization in science and engineering disciplines. Optimization is an intelligent method to explore the optimal solution among all obtainable ones of a specific problem [4][5]. The conventional approach for the multiobjective optimization problem is to scalarize the vector of multiobjective into a single objective by averaging the value returned by objective functions with a weight vector. Converting MOP into a single objective optimization problem allows a single-objective algorithm to be used more straightforwardly. Still, the obtained optimal solution is primarily dependent on the weight vector considered during the scalarization process. Moreover, a decision-maker has to know the problem in advance and provide a weight against each objective.

Furthermore, the decision-maker would be more interested in learning alternate solutions if obtainable. Algorithms are associated with such a predefined assumption, making them ineffective in solving today's compounded optimization problems. Solving a multiobjective optimization problem is one of such kind where a set of uniformly distributed solutions are required along the Pareto-optimal front in the search space.

Swarm-based optimization algorithm is a metaheuristic algorithm that works with a group of solutions and attempts to achieve the optimal solution in each iteration. It is often inspired by biological groups' sociality and has been widely applied for many real-world optimization problems to overcome the traditional optimization approach's restrictions. Such an intelligent system brings the advantages of being workable and straightforward to deal with different optimization problems. Additionally, it has the natural tendency to work with problem represented as the population of the optimal solution. Moreover, it takes place with the benefits of inherent movement, exploration, and exploitation, that lessen the possibility of entrapment into local optima. A number of Pareto-optimal sets can be obtained using a swarm-based optimization method for the multiobjective optimization problem.

The Salp swarm algorithm (SSA) is a new swarm-based metaheuristic method that imitates Salps flocking behavior in the ocean by making a chain. The SSA is similar to other evolutionary algorithms in many features, and it works proficiently for numerous real-world optimization problems. The flocking behavior of SSA can avoid entrapment of each solution into local optima up to some point due to its salp chain behavior. However, there are optimization problems where SSA cannot obtain a solution and easily trap into a local or deceptive optimum. The multiobjective optimization problem is one of such kind. The difficulty for SSA lies mainly due to a good search strategy for the multiobjective optimization problem. The original design of SSA is to save only one solution and update the positions based on the food source to obtain the best solution. However, there is no single best solution for the MOPs. It is required to obtain a set of well-distributed optimal solutions along the

Pareto-optimal front in the search space. Thus, there is a need to modify the SSA algorithms' original design to perform the overall searching process well with a balance of exploration and exploitation propensity to achieve the expected results.

The quantum-inspired algorithm is a new branch of study in the area of evolutionary computation. It is characterized by the particular principles of quantum physics such as uncertainty, superposition, interference, etc. The approach to merge and design the quantum-inspired algorithms for classical computers represents the solutions into quantum representation. The principles of quantum computing offer better diversity during the optimization process. The quantum search strategy intelligently guides the individuals toward the global optima by significantly improving convergence speed and solution efficiency. The variety can be derived from the representation model of the population. A probabilistic model of a linear superposition of states presents better characteristics of generating diversity in the population. Maintaining a good assortment in the population increases the searchability of the algorithm and resolves search stagnation. Sun et al. presented Quantum-behaved Particle Swarm Optimization (QPSO) to improve the performance of PSO, including good convergence and global search ability [6]. The approach guaranteed subjectively to discover a reasonable optimal solutions in the search space. The experimental results indicate that the QPSO works better than PSO and is a promising approach. Hence, in this study, we propose to employ quantum's inspiration to standard SSA for the same reason as QPSO for multiobjective optimization problems and compare it to MSSA and NSGA-II.

This article introduces a new presentation for the Multiobjective Quantum inspired Salp Swarm Algorithm (MQSSA) to ameliorate the overall performance of SSA. The approach is a hybrid of novel paradigms: SSA and Quantum Computing. Besides many other essential properties, this model can find a suitable solution faster using fewer individuals. This approach reduces the required number of estimation dramatically, which is a predominant effecting factor for solving the optimization problems. The Delta potential-well model (DPWM) representation of SSA in this paper enhances the convergence speed faster than traditional SSA. It maintains the population's diversity, preventing the population from stagnating in deceptive optima and increasing the algorithm's searchability. According to DPWM, if individual Salp in the SSA algorithm has quantum behavior, the algorithm is bound to work differently due to Heisenberg's uncertainty principle of quantum physics. To the best of our knowledge, a Quantum-inspired approach to improving the performance of SSA for the multiobjective optimization problems is introduced the first time. In this new approach, a simplified representation with DPWM integration to improve original SSA performance is introduced for MOP, making this algorithm more comfortable to understand and implement. The proposed algorithm's performance is evaluated on the multiobjective domain's complex benchmark problems in this work. A comparative study is performed to assess the performance with well-regarded algorithms MSSA and NSGA-II.

The rest of this paper is structured as follows: Standard Salp Swarm Algorithm (SSA) is presented in the next section, followed by the proposed algorithm with Delta potential-well model (DPWM). Experimental procedure followed by the empirical study and analysis, then a section discussing the results followed with conclusion.

9.2 SALP SWARM ALGORITHM

Salp Swarm Algorithm (SSA) is recently introduced swarm intelligence optimization algorithm. It is introduced by Mirjalili et al. in December 2017 [7]. After that, many types of research work for the enhancement and implementation of SSA are employed in engineering and technology, feature selection, and the multi-controller placement problem. A simpler version of SSA is introduced with a random search radius, in which the efficiency of the algorithm was improved [8]. Another particle-based SSA proposes balancing the exploration and exploitation propensity, which helps to boost the convergence speed and estimation accuracy [9]. Similarly, to ameliorate the searchability of the algorithm, the SSA is hybridized with the gravitational search algorithm [10]. Although the Salp Swarm Algorithm has been used broadly in many areas of engineering fields such as power system optimization and multilevel color image segmentation [11] [12], the recent literature shows lukewarm interest in the optimization problems where multiple scenarios are required to considered, that is, a multiobjective SSA (MS-MOSS) studied for sizing the photovoltaic system in which the number of PV modules and storage battery is pivotal and influences the soundness and cost of the system [13].

Figure 9.1: Salp chain in SSA.

Figure 9.1 shows a salp chain that is created by group of salps in deep ocean. According to the research, this chain follows the leader and follower pattern, where leaders' responsibility is to direct the chain toward food source and followers align with the leader in turn. For more details, refer to [7]. The mathematical representation and implementation of this chain categorize the Salps population into the front salps as leader, which position being updated based target and the rear salps as follower, which updates its position to align with adjacent individuals. This complete procedure is divided into three parts: Initialization, Define leaders, and Salps position updates.

9.2.1 INITIALIZATION

F in equation (9.1) represents an objective function of minimization problem:

$$F = f(x_1, x_2, x_3, x_4, x_5, ..., x_D) \tag{9.1}$$

D is the dimension in equation (9.1) and x_i is in constant boundaries such as $lb_i \leq x_i \leq ub_i$ ($i = 1, 2, 3, ..D$), where lb_i and ub_i are the lower and upper bounds of ith variable. Equation (9.2) calculates the initial position of each Salp for the population size N as the number of search agents, in which $rand(N, D)$ is a $N * D$ matrix with each element as a random number between 0 and 1:

$$X_{N*D} = rand(N, D) . * (ub - lb) + lb \tag{9.2}$$

9.2.2   LEADERS' SPECIFICATION

On initial position, each individual fitness is calculated using equation (9.1) and N fitness values obtained, respectively. Then the fitness values are sorted in ascending order and the minimum value from the vector, i.e., first position is regarding as the leader position, which is nearest to the global optimum. The rows of matrix X are rearranged based on sorted fitness values consequently, to appropriately distinguished between the leaders and followers salps. Mathematically, one from all N salps, the leader salp is constituted by row vector x^1 whose dimension is D. Therefore, the jth part of the leader salp represented as x_j^1 where $j = 1, 2, 3, ...D$. Similarly, the ith follower in dimension j^{th} represented as x_j^i where $i = 1, 2, 3, ...N$.

9.2.3   UPDATING POSITION

The leader's and followers' positions are updated as equations (9.3) and (9.4), respectively, the coefficient c_1 of the leader's equation is calculated using the mathematical equation (9.5). The position vector x_j^1 in the equation (9.3) represents the position of leaders in j^{th} dimension value and first individual with the best fitness value; similarly, F_j represents the position of the food source (F) in j^{th} dimension. The algorithm parameters c_2 and c_3 are evenly distributed random numbers between 0 and 1. The upper bound and lower bound of the search space are represented as ub_j and lb_j respectively in j^{th} dimension.

$$x_j^1 = \begin{array}{l} F_j + c_1 \left((ub_j - lb_j) c_2 + lb_j \right) c_3 \geq 0 \\ F_j - c_1 \left((ub_j - lb_j) c_2 + lb_j \right) c_3 \leq 0 \end{array} \tag{9.3}$$

$$x_j^i = \frac{1}{2} \left(x_j^i + x_j^{i-1} \right) \tag{9.4}$$

$$c_1 = 2e^{-\left(\frac{4t}{T} \right)^2} \tag{9.5}$$

This is important to note that updating leaders' position by determining the updating direction c_3 by equation (9.3), only related to individual toward global optimum and it has nothing to do with historical position of salps. However, equation (9.4) updates the followers position with a mechanism of adjacent salps, where j^{th} variable of the ith individual is x_j^i and the jth variable of the ith position of the nearby individual in front of the ith individual is x_j^{i-1}. Further in equation (9.5) c_1 is the most important parameter of the algorithm, which regulates the steps of position updates, which is defined with the current iteration t, maximum iteration T, and a natural constant e or Euler's number. From equation (9.5), it can be clearly observed that the value of c_1 is larger for the smaller number of iterations. Hence, in the initial iteration, the value of c_1 is for the exploration and at the later iterations for the exploitation of the search space.

9.2.4 RE-EVALUATION AND DECISION-MAKING

On the updated position, each individual fitness is calculated again using equation (9.1) and N fitness values obtained correspondingly. The leader position is revised again based on the updated fitness value to move the salp chain toward global optimum. The steps of iterations repeated for the maximum number of iterations to complete the optimization process.

9.3 PROPOSED MULTIOBJECTIVE QUANTUM-INSPIRED SALP SWARM ALGORITHM

Standard Salp Swarm Algorithm (SSA) has the superior ability to simplify the updating positions of salps by straightforward position update equations. However, SSA cannot produce expected results for the multiobjective optimization problems due to the same reason, as mentioned in Section 9.1. In this section, a new algorithm-based on SSA and Quantum Computing (QC) is presented. In QC, when considering the dynamics of particles and to avoid explosion with guaranteed convergence, particles must be inbound state during the run, that is, moving in a potential attraction field towards a center point [14]. There is potential field model such as Delta potential-well and Quantum oscillator to ascertain the particles movements in a bound state. In this paper, we employ the Delta potential-well model for the convergence of each particle having quantum behavior because particles' convergence is much faster in the quantum oscillator, leading to prematurity [6].

In this proposed algorithm, the algorithm's representation is changed based on the principle of DPWM and integrated into the standard SSA. The inspiration of the proposed MQSSA is based on a similar approach to QPSO; the aim is to improve the overall performance of SSA for multiobjective optimization problems. Further, this modification aims to adapt SSA and enhance the exploration ability and increase the algorithm's overall performance by an appropriate balance between exploration and exploitation tendency. This embedding approach of SSA and quantum computing is a promising strategy to improve the performance of SSA. One side, a DPWM representation known for increasing the speed of convergence and diversity in the population and on the other side integration with SSA to design a better search strategy increases the population's global searchability.

9.3.1   DELTA POTENTIAL-WELL MODEL FOR SSA

The Delta potential-well model describes using Dirac delta function as a theorized function in quantum physics. It approximately corresponds to the potential, which is zero throughout omitting a single point where it takes an infinite value. Each Salp (X^i) has a quantum state, and it can be formulated using the wavefunction $\Psi(x, t)$. From the analysis of the movements of salps toward a center point A_d, we assume that each Salp moves in a Delta potential well of search space for which the center point is A_d, that is, described as in equation (9.9).

For the simplicity of the presentation, considering in one-dimensional space with center point A, the Delta potential well can be represented as follows:

$$Z((x)) = -\gamma\delta(x-A) = -\gamma\delta(y) \tag{9.6}$$

where γ is a constant positive value and $\gamma\delta(y)$ denotes the Dirac delta function for y = (x A).

9.3.2   SALP POSITION MEASUREMENT

Multiobjective quantum-inspired SSA algorithm employs the Delta potential-well model representation from the quantum mechanics and considers salps movements towards a center point A_d as per equation (9.9), which is updated based on the lead position (F) stated in Section II as the best location. The potential field model and quantum oscillator in quantum physics ascertain the particles movements in a bound state due to the characteristics that each particle oscillate in a potential attraction field. The salps in the delta potential-well model must move in the bound state according to this strategy. To evaluate the fitness of each Salp, the position of the Salp needs to be calculated. However, only the probability of position for each Salp (X^i) can be learned from the probability density function $|\Psi(x, t)|^2$, which depends on the potential field the particle lies into. The probability of position indicates that a salp appears at position x relative to a point A_d. Hence, it is required to measure salp position using the collapsing technique, that is, transformation from a quantum state to the classical form. The Monte Carlo method can reproduce this process of measurement [15][16].

In this paper, the following iterative equations were used to measure each salp according to the Monte Carlo method. As stated, salps moved round and swayed toward A_d with its kinetic energy declining to zero. With this characteristic in the modified algorithm, the salps are inbound states and avoid explosion with guaranteed convergence.

$$X_{j+1}^k = \begin{array}{l} A_d + B_l \times |BestMean_l - X_j^k|log(\frac{r_d}{u_d}) \quad c_3 > 0 \\ A_d - B_l \times |BestMean_l - X_j^k|log(\frac{r_d}{u_d}) \quad c_3 < 0 \end{array} \tag{9.7}$$

where d is the current dimension and l is the current iteration.

We assumed here the first iteration $l = 0$ with maximum iteration size L. The r_d and u_d are random numbers in the range [0, 1] in $d - dimension$. B_l describes equation (9.8), a contraction-expansion coefficient that is adjusted dynamically during the optimization process. It means gradually decreasing or increasing iteration wise according to individual salp's convergence speed and algorithm performance. A_d described as in equation (9.9), a center point for salps to move toward locally. Also known as the learning inclination point for salps to oscillate around. $BestMean_l$ describes mathematically as equation (9.10), mean of the best position. X_j^k is the kth salp in $j - dimension$ and X_{j+1}^k is the new position of salp.

$$\beta = \left(\frac{0.59 * (L - l)}{l + 0.59} \right)$$

where the current iteration is l, and L is the maximum number of iterations

$$B_l = (\beta + (1 - \beta)) \tag{9.8}$$

$$A_d = \frac{\left(r_{1d} * X_j^k + r_{2d} * Leader_j \right)}{r_{1d} + r_{1d}} \tag{9.9}$$

where r_{1d} and r_{2d} are random numbers in the range [0, 1]. $Leader_j$ is the leader position and represented as the best location.

$$BestMean_l = \frac{1}{N} \sum_{j=1}^{d} lead_j (l) \tag{9.10}$$

where N is the maximum number of populations.

9.3.3 THE NEW ALGORITHM BEHAVIOR

This hybrid approach with quantum computing to enhance the standard SSA performance boosts the overall performance and maintains individual diversity for the multiobjective optimization problems. In which it produces a set of well-distributed Pareto-optimal solutions. Preserving diversity is extremely important for such optimization problems as it is often entrapment into non-distributed solutions. The searching tendency of evolutionary algorithms is determined based on the population diversity. It means lower exploration ability in the populations identical elements.

In MQSSA, equation (9.7) updates the salps position and produces new solutions. Pseudo-code for the MQSSA is presented in Algorithm 1 of Figure 9.2. After the initialization phase of the population and first approximation calculation in steps 2 and 3, the algorithm takes up into the optimization process' main loop. This loop executes for a maximum number of MAX generations and is formed alongside to perform various tasks. First, the control parameters for the standard SSA and MQSSA calculated according to equation (9.5) and (9.8), respectively, followed by calculating the objective function values for each Salp and identify other than dominated ones at step 6. Step 7 handles the repository to update for the obtained non-dominated solutions. The other than dominated solutions to be added in repository if it is not full. At this stage, the non-dominated solutions are compared with available solutions in the repository to keep only other than dominated solutions in the repository. After that, it also deletes the solutions from the crowded region to maintain the number of solutions. For this, it first ranked the solutions and then selected using the roulette wheel technique.

Step 10 executes only once at first iteration and generates a salp chain where a division of population happens into leaders and followers. After that, this algorithm is used in step 8, followed by evaluating solutions for the coverage and updating the search boundary. The process repeats for MAX generation, and in each generation, it produces Pareto optimal solutions, which is a set of uniformly distributed solutions in the search space along the estimated Pareto optimal front. At step 8, the interference process and proposed quantum-based equations for SSA are executed to generate new solutions. This process consists of updating the individual position using equation (9.7), calculating the contraction-expansion coefficient, evaluating the converging points and best mean along with the fitness of salps. At step 7, the other than dominated solutions to be put into an archive if the archive has an empty space, i.e., its not full. At this stage, the other than dominated solutions are compared with available solutions in the repository to keep only other than dominated obtained solutions in the repository. After that, at step 11 CoverageSelection() function used to delete the solutions from the crowded region to maintain the number of solutions. For this, it first ranked all the solutions and then selected using an intelligent roulette wheel technique. At step 14, the algorithm returns the obtained Pareto-optimal set as the best solution along with the first approximation.

9.4 EXPERIMENTAL PROCEDURE

9.4.1 COMPUTING ENVIRONMENT

The source code of MQSSA is developed in MATLAB R2017b, and the experimental computer environment for the simulation is used as follows:

1. Intel Core i7-3520M CPU @ 2.90 GHz
2. 16 GB of RAM
3. macOS Catalina v.10.15.7 operating system

9.4.2 PERFORMANCE ASSESSMENT METRICS

To evaluate the performance of the MQSSA algorithm, two metrics are used in this paper: Pareto Sets Closeness (PSC) and Inverted Generational Distance metric (IGD) [17]. PSC assesses the similarity between the approximated PF and the obtained Pareto-optimal set. The IGD indicator assesses diversity and convergence both for the solutions obtained in the search space. The more considerable value of PSC means a well-distributed obtained solution in the search space. The lower value of IGD means that the diversity and convergence of the obtained solutions in search space are adequate.

The IGD can be defined as an average Euclidean distance between the approximated Pareto-optimal set and obtained Pareto-optimal set in the search space. Let U^*

---

1.   **procedure** MQSSA (Max, ITR, lb, ub, dim, fObj)

2.   **Initialization:** pop, param, Best, Fitness, subPop
3.   **Fitness** calculation and first approximation

4.   **while** ITR ≤ Max **do**   *Main Loop
5.   Control parameters calculation
6.   Non-dominatedSalps ()
7.   HandleRepository ()

8.   **if** ITR ≥ 1 **then**
                   Update positions equation (7)
                   Parameters updates equation (8), (9), (10)

9.   **else**

10.   Build salp chain using equation (3), (4)

11.   CoverageSelection ()
12.   UpdateBounds ()
13. **end** while

14. Best Solution   *Returns the Obtained Pareto-set

Figure 9.2: The Algorithm MQSSA.

---

denotes a set of uniformly distributed exact Pareto-optimal location and H represents a set of obtained Pareto-optimal solutions of the final population. The IGD can be computed as the average distance from U^* to H in the decision space using the below equation:

$$IGD\ (H,\ U^*) = \frac{\Sigma_{u \varepsilon U^*}\ \min d(u,\ H)}{|U_*|}$$

where u represents an exact point in U^*, $\min d\ (u,\ H)$ is the minimum Euclidean distance between u and the points in H.

Similarly, the PSC value can be calculated using the below equation, which indicates the closeness between the approximated PSs and obtains PSs.

$$PSC = \frac{CR}{IGD}$$

CR shows the overlapping ratio between the approximated Pareto-optimal set and the obtained Pareto-optimal set, and IGD is the inverted generational distance in the search space. A higher PSC value means the obtained solutions are well distributed in the search space and is adequate.

Table 9.1

Benchmark Problems Used in This Study

Function	Objective Function	DIM	Variable Bounds	Characteristics of the PF
SCH1	$f_1(x) = x^2$ $f_2(x) = (x - 2)^2$	1	$x \in [-10^3, 10^3]$	Convex
ZDT1	$f_1(x) = x_1$ $f_2(x) = g(x)\left[1 - \sqrt{x_1/g(x)}\right]$ $g(x) = 1 + 9\left(\sum_{i=2}^{n} x_i\right)/(n-1)$	30	$x_i \in [0, 1]$	Convex
ZDT2	$f_1(x) = x_1$ $f_2(x) = g(x)\left[1 - \left(x_1/g(x)\right)^2\right]$ $g(x) = 1 + 9\left(\sum_{i=2}^{n} x_i\right)/(n-1)$	30	$x_i \in [0, 1]$	Nonconvex
ZDT3	$f_1(x) = x_1$ $f_2(x) = g(x)\left(1 - \sqrt{x_1/g(x)}\right.$ $\left. - x_1 sin(10\pi x_1)/g(x)\right)$ $g(x) = 1 + 9\left(\sum_{i=2}^{n} x_i\right)/(n-1)$	30	$x_i \in [0, 1]$	Convex Disconnect
ZDT4	$f_1(x) = x_1$ $f_2(x) = g(x)\left(1 - \sqrt{x_1/g(x)}\right)$ $g(x) = 1 + 10(n-1) + \sum_{i=2}^{n} [x_i^2 - 10cos(4\pi x_i)]$	10	$x_i \in [0, 1]$	Nonconvex
FON	$f_1(x) = 1 - exp\left(-\sum_{i=1}^{3}\left(x_i - \frac{1}{\sqrt{3}}\right)^2\right)$ $f_2(x) = 1 - exp\left(-\sum_{i=1}^{3}\left(x_i + \frac{1}{\sqrt{3}}\right)^2\right)$	3	$x_i \in [-4, 4]$	Nonconvex
KUR	$f_1(x) = \sum_{i=1}^{n-1}\left(-10exp\left(-0.2\sqrt{x_i^2 + x_{i+1}^2}\right)\right)$ $f_2(x) = \sum_{i=1}^{n}\left(|x_i|^a + 5sin(x_i^3)\right)$	3	$x_i \in [-5, 5]$	Nonconvex

9.4.3   MULTIOBJECTIVE BENCHMARK PROBLEMS

The MQSSA algorithm's performance is evaluated on the seven challenging well-known multiobjective optimization benchmark problems whose Pareto-optimal front is convex-shaped and concave-shaped. These multiobjective optimization benchmark functions are presented in Table 9.1 with the name of the function, dimension, ranges of variables, and PF characteristics. The benchmark functions are divided into functions according to the features of their shape of Pareto-optimal front, i.e. convex, concave, and convex disconnected. Benchmark functions SCH1 and ZDT1 are convex-shaped, whose surface is curved like the exterior of a circle. Functions ZDT2, ZDT4, FON, and KUR are concave-shaped, whose surface is curved inward like the circle's interior. Function ZDT3 is a convex disconnected problem, where the Pareto-optimal front is in the shape of convex but with isolated regions. These Pareto fronts are regular in many real-world optimization problems and are very difficult to be set on by the optimization approaches. All these benchmark problems have two objectives, with having no constraints.

Table 9.2

Statistical Results of Pareto Sets Proximity (PSP)

	MQSSA		MSSA		NSGA-II	
	Average	Std. Dev.	Average	Std. Dev.	Average	Std. Dev.
SCH1	9.49E-03	1.93E-03	2.64E-03	2.78E-03	9.19E-03	1.59E-03
ZDT1	5.36E-02	3.38E-02	1.88E-02	2.97E-02	0.00E+00	0.00E+00
ZDT2	4.34E-04	0.00E+00	0.00E+00	0.00E+00	0.00E+00	0.00E+00
ZDT3	7.21E-02	3.87E-02	4.02E-02	4.29E-02	0.00E+00	0.00E+00
ZDT4	1.65E-03	2.44E-03	1.96E-04	7.57E-04	2.47E-03	3.57E-03
FON	1.64E+00	4.55E-01	1.37E-02	3.79E-02	1.81E+00	2.38E-01
KUR	1.58E-02	7.03E-03	9.80E-03	8.15E-03	1.70E-02	6.00E-03

9.4.4   EVALUATING METHOD AND ALGORITHMS PARAMETERS

The MQSSA requires five parameters: Max-Iteration, Current-Iteration, lower bound, upper bound, dimension, and objective function reference. All the necessary parameters for the algorithm are being managed using the other functions such as the number of salps, best solution, etc. This paper performed a comparative study of MQSSA with MSSA [7] and NSGA-II [18].

For a fair performance comparison, the following parameters are used in this experiment:

1. Population size 60
2. Maximum number of generations 1000 for all the test functions
3. The same number of function evaluation is considered for all the test problems.
4. For NSGA-II, Crossover = 0.7, Mutation = 0.4 as percentage, and mu=0.02 as mutation rate.

The average results of 30 independent runs and average mean, the standard deviation of metrics PSP and IGD are summarized in Tables 9.2 and 9.3. The average norm indicates how the MQSSA performs on average, and the standard deviation shows how stable it is during all the runs. The experimental results demonstrate that the overall performance of MQSSA is competitive as compared with other approaches.

9.5   EXPERIMENTS AND DISCUSSION

The MQSSA is executed and evaluated for the multiobjective optimization benchmark problems, and quantitative results are calculated using the PSP and IGDX of each trial. They are recorded in the form of average and standard deviation (SD), presented in Tables 9.2 and 9.3 as results for the analysis. Also, the obtained solution as true Pareto optimal set is shown in Figures 9.3–9.9. The maximum size of repository is set to 100 for MQSSA. In this paper, the two highly regarded multiobjective

Table 9.3

Statistical Results of Inverted Generational Distance (IGD)

	MQSSA		MSSA		NSGA-II	
	Average	Std. Dev.	Average	Std. Dev.	Average	Std. Dev.
SCH1	1.28E+01	2.24E+00	1.45E+01	2.05E+00	1.31E+01	2.02E+00
ZDT1	3.11E+00	1.46E-01	3.43E+00	2.97E-01	6.60E+01	2.51E+00
ZDT2	4.47E+00	2.20E-01	4.50E+00	2.06E-01	9.02E+01	3.56E+00
ZDT3	3.10E+00	1.77E-01	3.40E+00	3.80E-01	4.87E+01	8.51E+00
ZDT4	1.75E+01	1.22E+00	1.83E+01	1.25E+00	1.74E+01	1.18E+00
FON	5.01E-01	7.26E-03	5.36E-01	3.97E-02	4.97E-01	1.08E-02
KUR	1.17E+01	6.11E-01	1.23E+01	6.51E-01	1.18E+01	5.67E-01

optimization algorithms in this domain are selected for the results endorsement and for the comparative study: MSSA and NSGA-II.

Tables 9.2 and 9.3 indicate that the MQSSA algorithm remarkably performs better than MSSA and NSGA-II on most of the ZDT functions. Further, when examined, the obtained PF in Figures 9.4–9.6 indicates that MQSSA shows a superior convergence than MSSA and NSGA-II for the ZDT benchmark functions. The obtained solutions are well distributed uniformly for MQSSA algorithm, which means the algorithm coverage is high. A space in midway of the obtained Pareto optimal set by MSSA and NSGA-II in some of the ZDT problems indicates that how the coverage of the algorithm is negatively impacted. The SCH1 and ZDT1 functions are convex-shaped as indicated by the Pareto optimal front. Hence, the coverage and convergence propensity of the algorithms can be benchmarked. The obtained Pareto-optimal set in Figures 9.3 and 9.4 indicate that MQSSA performs anew better than MSSA and NSGA-II algorithms. The coverage is deficient for the NSGA-II on ZDT2 function.

It looks that the convergence and coverage of MQSSA are better than MSSA. Further, it can be observed that the MSSA algorithm find a gap in the obtained Pareto optimal set, but the obtained solutions are well distributed in rest of the true Pareto-optimal set. Additionally, the problems ZDT2 and ZDT4 shown to be Pareto optimal front of type concave-shaped, which is invariably difficult for the algorithm designed based on aggregation approach. But, the obtained results indicates that MQSSA can efficiently approximate these functions' right front with exceedingly good coverage and convergence. When it is compared between MQSSA and NSG-II in Figures 9.4–9.6, results indicate that the coverage and convergence of MQSSA are better and keen.

Tables 9.2 and 9.3 and Figure 9.6 analysis of ZDT3 function has Pareto optimal set with isolated regions. Such kind of Pareto-optimal set is usual in the real-world optimization problems. It is very difficult for the algorithms to obtain the Pareto optimal set of such problems. There is a highly possibility that the algorithm is failed to obtain the Pareto-optimal set in all the separated areas and pin down in one of the region. The comparison outcome on ZDT3 and other previously discussed ZDT problems indicate that the achievement of NSGA-II is comparatively low, the coverage is good, but the convergence is inferior. The obtained Pareto optimal set of several isolated regions is far away from the true Pareto-optimal front. In the view of obtained Pareto-optimal front by MQSSA and MSSA, it is evident that MQSSA performs comparatively better than MSSA with regard to coverage and convergence. These outcomes exemplify that MQSSA can successfully discover all the isolated regions of the Pareto optimal front with well-distributed solutions in all the areas.

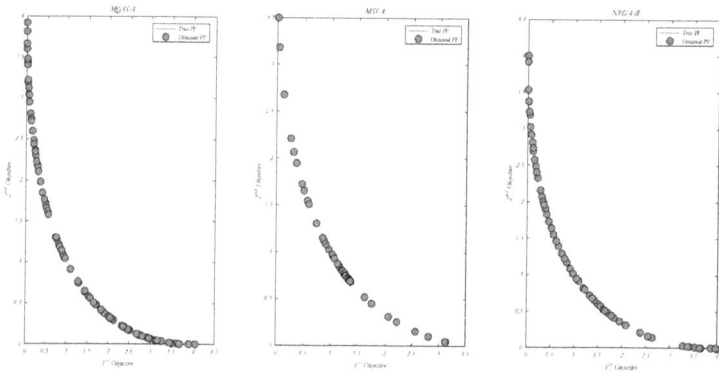

Figure 9.3: SCH1 Pareto front obtained by MQSSA, NSGA-II, and MSSA.

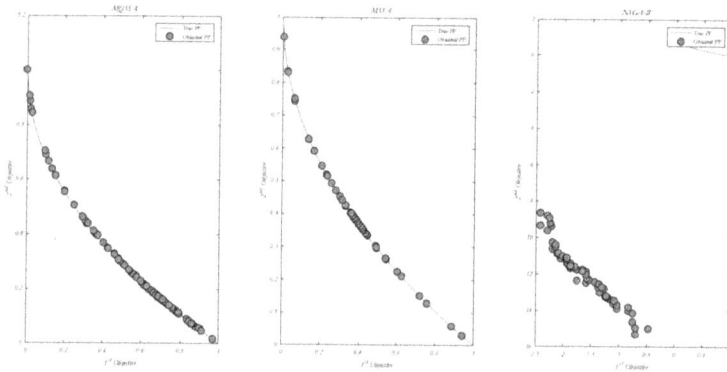

Figure 9.4: ZDT1 Pareto front obtained by MQSSA, NSGA-II, and MSSA.

The experimental results and above discussion prove that MQSSA can estimate Pareto optimal front of type concave and convex shaped with a reasonable coverage and convergence on four ZDT series benchmark problems: ZDT1, ZDT2, ZDT3, and ZDT4.

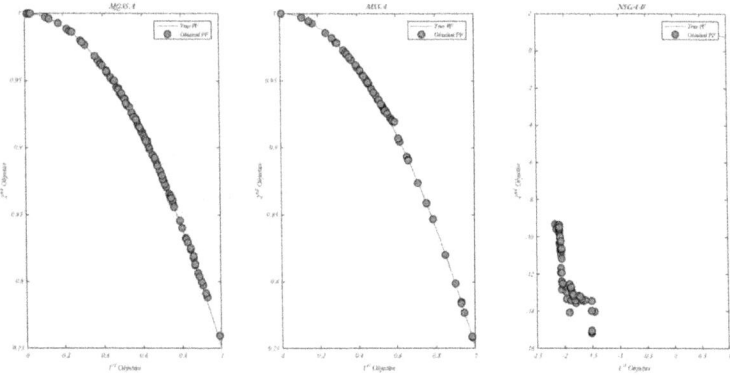

Figure 9.5: ZDT2 obtained Pareto front by MQSSA, NSGA-II, and MSSA.

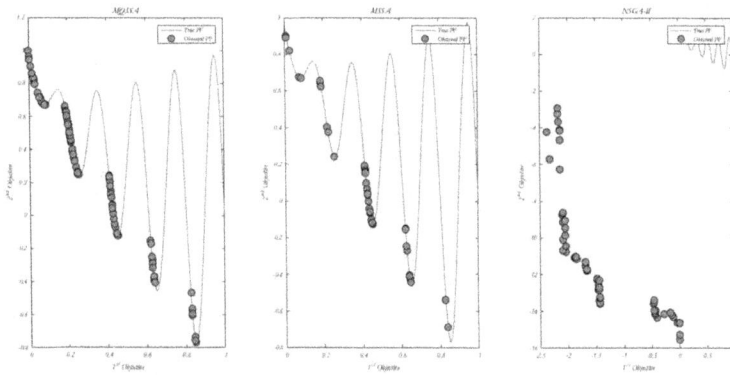

Figure 9.6: ZDT3 Pareto-front obtained by MQSSA, NSGA-II, and MSSA.

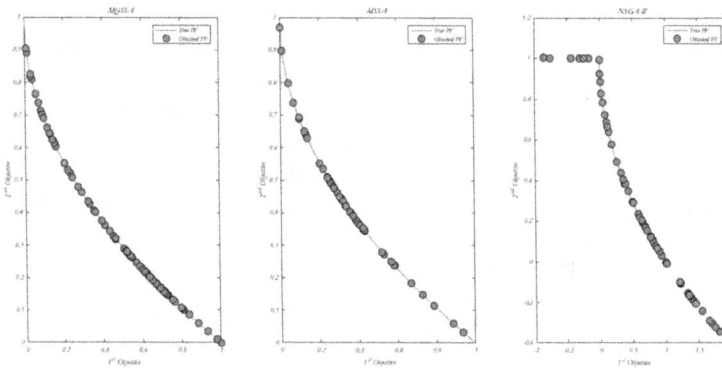

Figure 9.7: ZDT4 Pareto-front obtained by MQSSA, NSGA-II, and MSSA.

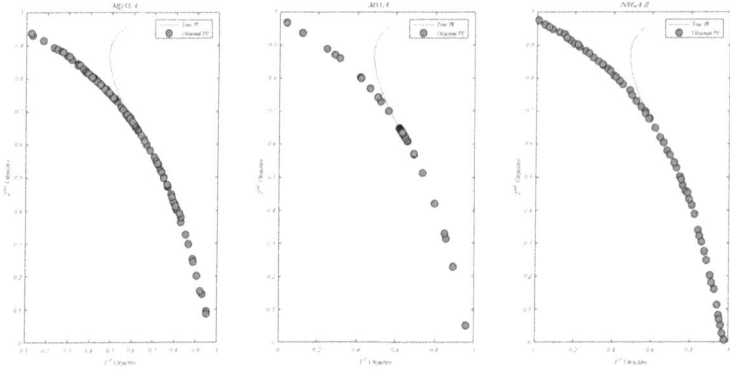

Figure 9.8: FON obtained Pareto-front by MQSSA, NSGA-II, and MSSA.

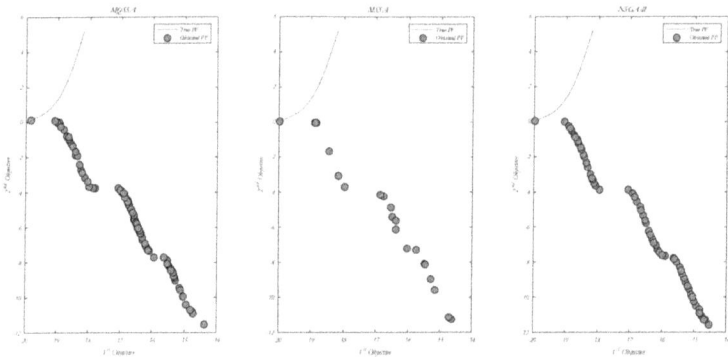

Figure 9.9: KUR obtained Pareto front by MQSSA, NSGA-II, and MSSA.

Further study on FON and KUR, which is equipped with nonconvex-shaped, can benchmark the convergence and coverage algorithm. The analysis outcome of Figures 9.8–9.9 shows that all three algorithms, MQSSA, MSSA, and NSGA-II, are unable to converge for KURs Pareto optimal front. Most of the obtained solutions are far from the Pareto-optimal front. However, the results from Tables 9.2–9.3 show that MQSSA performs slightly better. Inspecting the fronts of MQSSA, NSGA-II, and MSSA for FON shows that MQSSA and NSGA-II provide better convergence. However, the results from Tables 9.2 and 9.3 show NSGA-II performs slightly better.

These results exemplify that MQSSA successfully guide salp chain with regard to separate regions of true Pareto optimal front and is competitive as compared with other approaches.

9.6   CONCLUSION

This chapter presents a novel Multiobjective Quantum-inspired Salp Swarm Algorithm (MQSSA) with Delta potential-well model presentation, which is a better alternative than the binary presentation for the multiobjective optimization problems. The proposed approach is evaluated on several multiobjective optimization benchmark problems having convex-shaped and concave-shaped Pareto-optimal front. It shows a favorable outcome, with better performance than other well-regarded algorithms in the multiobjective domain. Further investigations would be required to evaluate the robustness of MASSA on solving different kinds of optimization problems.

Besides, the experimental study showed that the proposed algorithm, MQSSA:

1. Has the ability to achieve the Pareto-optimal front for the large dimensional multiobjective optimization problems.
2. Excellent speed as compared to the traditional SSA.
3. An appropriate balance between exploitation and exploration propensities.
4. Appropriate convergence rate and good coverage.

Future study would include applying the proposed algorithm to solve more benchmark problems and real-world optimization problems from different domains. It would also be interesting to study the relationship and the differences between this algorithm and other optimization approaches, such as several improved PSO and other swarm intelligence techniques.

REFERENCES

1. H. Li, K. Deb, and Q. Zhang, "Variable-length Pareto optimization via decomposition-based evolutionary multiobjective algorithm," IEEE Trans. Evol. Comput., vol. 23, no. 6, pp. 987–999, Dec. 2019.
2. K. Li, R. Chen, G. Fu, and X. Yao, "Two-archive evolutionary algorithm for constrained multiobjective optimization," IEEE Trans. Evol. Comput., vol. 23, no. 2, pp. 303–315, Apr. 2019.
3. Y. Sun, B. Xue, M. Zhang, and G.G. Yen, "A new two-stage evolutionary algorithm for many-objective optimization," IEEE Trans. Evol. Comput., vol. 23, no. 5, pp. 748–761, Oct. 2019.
4. X. Yang, "Review of meta-heuristics and generalized evolutionary walk algorithm", Int. J Bio-Inspir. Com., vol. 3, no. 2, pp. 77–84, 2011.
5. Z. Michalewicz, Genetic algorithms + data structures = evolution programs (2nd, extended ed.). New York, NY: Springer-Verlag New York, Inc., 1994.
6. J. Sun, B. Feng, and W. Xu, "Particle swarm optimization with particles having quantum behavior," in Proc. Congr. Evol. Comput., vol. 1, pp. 325–331, 2004.
7. S. Mirjalili, A.H. Gandomi, S.Z. Mirjalili, S. Saremi, H. Faris, S.M. Mirjalili, "Salp swarm algorithm: A bio-inspired optimizer for engineering design problems," Adv. Eng. Softw., vol. 114, pp. 163–191, 2017.
8. B. Xiao, R. Wang, Y. Xu, J. Wang, W. Song, and Y. Deng, "Simplified Salp Swarm Algorithm," 2019 IEEE International Conference on Artificial Intelligence and Computer Applications (ICAICA), Dalian, China, 2019, pp. 226–230, doi: 10.1109/ICAICA.2019.8873515.
9. B. Xiao, R. Wang, Y., Xu, et al. "Salp Swarm Algorithm based on Particle-best", In: IEEE 3rd Information Technology, Networking, Electronic and Automation Control Conference (ITNEC). Chengdu, pp. 1383–1387, 2019.

10. S. Li, Y. Yu, D. Sugiyama, D., et al. "A Hybrid Salp Swarm Algorithm With Gravitational Search Mechanism," In: 5th IEEE International Conference on Cloud Computing and Intelligence Systems (CCIS), Nanjing, pp. 257–261, 2018.

11. S. Ekinci and B. Hekimoglu, "Parameter optimization of power system stabilizer via Salp Swarm algorithm," In: 5th International Conference on Electrical and Electronic Engineering (ICEEE), Istanbul, pp. 143–147, 2018.

12. Z. Xing and H. Jia. "Multilevel Color Image Segmentation Based on GLCM and Improved Salp Swarm Algorithm," IEEE Access, vol. 7, pp. 37672–37690, 2019.

13. H.M. Ridha, C. Gomes, H. Hizam, and S. Mirjalili. "Multiple scenarios multiobjective salp swarm optimization for sizing of standalone photovoltaic system," Renewable Energy, Elsevier, vol. 153(C), pp. 1330–1345, 2020.

14. M. Clerc and J. Kennedy, "The Particle Swarm: Explosion, stability and convergence in a multi-dimensional complex space", IEEE Trans. Evol. Comput., 6: 58–73, 2002.

15. J. Liu, W. Xu, and J. Sun, "Quantum behaved particle swarm optimization with mutation operator", In: Proc. of IEEE International Conference on Tools with Artificial Intelligence, pp. 240–244, 2005.

16. J. Sun et al, "A global search strategy of quantum-behaved particle swarm optimization", IEEE Conference on Cybernetics and Intelligent Systems, pp 111–116, 2004.

17. G. Li, L. Yan, and B. Qu, "Multiobjective particle swarm optimization based on Gaussian sampling," IEEE Access, vol. 8, pp. 209717–209737, 2020, doi: 10.1109/ACCESS.2020.3038497.

18. K. Deb, S. Agrawal, A. Pratap, and T. Meyarivan. A Fast Elitist Non-dominated Sorting Genetic Algorithm for Multiobjective Optimization: NSGA-II. In: Schoenauer M. et al. (eds) Parallel Problem Solving from Nature PPSN VI. PPSN 2000. Lecture Notes in Computer Science, vol. 1917. Springer, Berlin, Heidelberg, 2000. https://doi.org/10.1007/3-540-45356-383.

10 Quantum-Inspired Multi-Objective NSGA-II Algorithm for Automatic Clustering of Gray Scale Images

10.1 INTRODUCTION

Clustering [1][2][3] is a process of partitioning a heterogeneous dataset into some groups of homogeneous data points or elements. It can be considered as a challenging task to generate appropriate number of cluster from a given dataset due to lack of proper knowledge of the dataset. By addressing the problem of automatic clustering, several research works have been done so far, some of them are presented in [4][5][6][7]. While considering the purpose of clustering, sometimes the clustering algorithms provide good results with one type of datasets but not be able to provide good results with other types of dataset, which yields more challenges in the field of automatic clustering [12]. Though, there exist many *automatic clustering* algorithms [8][9][10][11], but those algorithms have focused only on optimizing a single objective, whereas in real-world scenario, many problems may have more than one objective which need to be taken for consideration [13]. In this aspect, the requirement of multi-objective optimization has been realized by which multiple objectives which are not only similar but even possibly conflicting can be tackled properly.

Nowadays, *multi-objective* optimization algorithms are becoming popular for their capability of searching a highly complex search space. In the last decade, researchers have developed many nature-inspired multi-objective optimization algorithms, which include non-dominated sorting GA (NSGA-II) [14][15], Pareto envelope-based selection algorithm (PESAII) [16], Strength Pareto Evolutionary Algorithm (SPEA) [17], and its improved version, SPEA2 [18]. The overview and applicability of the multi-objective algorithms for clustering are presented by Maulik et al. in [19]. Zhou et al. addressed the basic principles, advancements, and applications of multi-objective algorithms to solve several real-world optimization problems in [20].

In recent years, the concepts of quantum computing [21] are being incorporated with the evolutionary algorithms for effectively exploring the search space for *multi-objective* optimization problems. Quantum-inspired evolutionary algorithms have been developed for performing quasi quantum operations on classical computers by implementing the concept of quantum mechanical phenomena. Few research works on quantum-inspired evolutionary algorithms are available in [17][23][24][25][26][27][28]. Though, the *quantum-inspired* evolutionary algorithms can efficiently explore and exploit search space for a global optimal solution, still there remains a challenge to handle the problem of multi-objective optimization. It can be handled by the help of quantum-inspired algorithms to improve the proximity of the non-dominated Pareto optimal front and by employing advantages of quantum-inspired algorithms to preserve the diversity.

In this paper, NSGA-II [15] has been taken as the base algorithm for implementing the proposed work as NSGA-II [15] uses a strong elitist mechanism for maintaining the diversity of the search space efficiently. It also uses *non-dominated* sorting and *crowding distance* assignment for identifying the non-dominated Pareto optimal front easily. With the help of quantum-inspired mechanism incorporated with NSGA-II [15], it is possible to increase the efficiency of the existing elitism method as, for a single solution, there exists multiple observations of qubit individuals, which allow a local search in the neighborhood of the non-dominated solutions. Furthermore, the possibility of losing high quality individuals can be reduced by selecting the best qubit individuals in every generation. The convergence and preservation of diversity being the main issues under scrutiny, the proposed algorithm is expected to improve the performance of its counterpart when used for automatic clustering. In this paper, an effort has been made to identify the optimal number of clusters from a gray scale image on the run by implementing a quantum-inspired algorithm, viz., Quantum-Inspired Multi-objective NSGA-II (QIMONSGA-II), which basically performs quasi quantum computation and optimizes two different objectives, viz., CS-Measure (CSM) [7] and DB index [30] simultaneously. All the experiments have been performed over six Berkeley [31] gray scale images of different sizes.

The rest of the paper is organized as follows: The fundamental concepts of quantum computing are presented in Section 10.2. The computational processes of both the objectives functions are presented in Section 10.3. Section 10.4 presents a brief introduction regarding the multi-objective optimization along with the basic steps of NSGA-II algorithm for automatic clustering of gray scale images. Section 10.5 provides an elaborate description of the proposed work followed by Section 10.6 with the discussion on experimental results and analysis. Finally, the paper is concluded with a conclusion and future guidance in Section 10.7.

10.2   QUANTUM COMPUTING FUNDAMENTAL

In recent years, quantum computing, a new computational paradigm has been invented by Deutsch and Feynman [21][32] in 1982, which utilizes the fundamental principles, viz., quantum bits or qubits, superposition, and entanglement, of quantum mechanics for performing any calculation. Unlike classical computer, a quantum computer uses quantum bits *qubit* as a basic computational unit. A single qubit may have any one of the three states, viz., $|1\rangle$, $|0\rangle$ or any superposition state of these two states, at the same time [21]. The state of a qubit can be represented as

$$|\Psi\rangle = \alpha\,|0\rangle + \beta\,|1\rangle \tag{10.1}$$

where α and β are complex numbers. The probability amplitudes α^2 and β^2 specify the probabilities of the state to be in $|0\rangle$ or $|1\rangle$, respectively. The superposition state $|\Psi\rangle$ can be realized by the following:

$$|\Psi\rangle = \begin{cases} |0\rangle, & \text{if } |\alpha|^2 > |\beta|^2 \\ |1\rangle, & \text{Otherwise} \end{cases} \tag{10.2}$$

where α^2 and β^2 should always satisfy the following equation:

$$|\alpha|^2 + |\beta|^2 = 1 \tag{10.3}$$

10.3 COMPUTING THE OBJECTIVES

Two different cluster validity indices, viz., CS-Measure (CSM) [7] and DB index [30], have been used as the optimization functions, which need to be optimize simultaneously.

10.3.1 CS-MEASURE (CSM) INDEX

In 2004, Chou *et al.* proposed a *cluster validity index* , viz., CS-Measure (CSM) [7] index, which can efficiently handle the clusters of different densities and/or sizes by computing the ratio of sum of within cluster scatter and between cluster separation. It initially identifies the cluster centroid C_i from a data set by averaging the data points. The cluster centroid C_i can be created by the following equation.

$$C_i = \frac{1}{N_i} \sum_{D_i \in D_{Si}} D_i \tag{10.4}$$

where the total number of data points (D_i) is represented by N_i in the ith cluster of the data set D_S. Now, the mathematical representation for the CSM index can be defined by the following equation in which the data set D_S contains N_C number of clusters.

$$CSM = \frac{\frac{1}{N_C} \sum_{i=1}^{N_C} \left[\frac{1}{|N_i|} \sum_{D_i \in D_{Si}} \max_{D_{mx} \in D_{Si}} \left\{ D_{iff}(D_i, D_{mx}) \right\} \right]}{\frac{1}{N_C} \sum_{i=1}^{N_C} \left[\min_{j \in N_C, j \neq i} \left\{ D_{iff}(C_i, C_j) \right\} \right]} \tag{10.5}$$

$$\therefore CSM = \frac{\sum_{i=1}^{N_C} \left[\frac{1}{|N_i|} \sum_{D_i \in D_{Si}} \max_{D_{mx} \in D_{Si}} \left\{ D_{iff}(D_i, D_{mx}) \right\} \right]}{\sum_{i=1}^{N_C} \left[\min_{j \in N_C, j \neq i} \left\{ D_{iff}(C_i, C_j) \right\} \right]} \tag{10.6}$$

where the difference between any two data points D_i and D_j is defined as $D_{iff}(D_i, D_j)$. The optimal result is achieved for a minimum value of CSM index [7].

10.3.2 DAVIES–BOULDIN (DB) INDEX

In the year 1979, David L. Davies and Donald W. Bouldin proposed a cluster validity index, viz., Davies–Bouldin (DB) index [30]. It computes a similarity measure R_{ij} between the clusters C_i and C_j, which is basically defined based on a measure of dispersion of a cluster C_i. Mathematically, the index R_{ij} is defined by the following equation.

$$R_{ij} = \frac{S_i + S_j}{D_{ij}} \tag{10.7}$$

where the dispersion measure of the ith cluster is represented by S_i as follows:

$$S_i = \left[\frac{1}{N_i} \sum_{l=1}^{N_i} \left\| X_l^{(i)} - Z_i \right\|^2 \right]^{\frac{1}{2}} \tag{10.8}$$

where the ith cluster center of a cluster C_i is represented by Z_i and the total number of objects is represented by N_i with all $X_l^{(i)}$ belonging to C_i. The distance between clusters C_i and C_j is measured in terms of the cluster dissimilarity measure D_{ij}, which can be defined as follows:

$$D_{ij} \left\| Z_i - Z_j \right\| \tag{10.9}$$

Finally, the Davies Bouldin (DB) index [30] is defined by the following equation.

$$DB = \frac{1}{N_c} \sum_{i=1}^{N_c} R_i \tag{10.10}$$

where $R_i = \max\limits_{j=1,2,...,N_c, i \neq j} (R_{ij}), i = 1, 2, ..., N_c$. While computing R_{ij}, the following constrains should be satisfied.

1. $R_{ij} \geq 0$
2. $R_{ij} = R_{ji}$
3. if $S_i = S_j = 0$ then $R_{ij} = 0$
4. if $S_j = S_k$ and $D_{ij} < D_{ik}$ then $R_{ij} > R_{ik}$
5. if $S_j > S_k$ and $D_{ij} = D_{ik}$ then $R_{ij} > R_{ik}$

The optimal result is achieved by minimizing the value of DB index [30].

10.4 MULTI-OBJECTIVE OPTIMIZATION

Multi-objective optimization is a process of identifying a set of solutions, called the *Pareto optimal* set, by optimizing more than one objective functions simultaneously [33]. In many real-life problems, there exists no single optimal solution for a given problem but may have a set of possible solutions of equivalent quality [13]. Such problems are addressed by *multi-objective* optimization algorithms by simultaneously optimizing the multiple and possibly competing objective functions. Unlike single-objective optimization problem, the goodness of a solution in a multi-objective optimization problem is determined by the dominance of the solutions.

The multi-objective optimization can be formally stated as [33] follows: In general, for a multi-objective optimization problem, M number of objectives are need to be optimized simultaneously by the decision variable vector $\bar{x}^* = [x_1^*, x_2^*, ..., x_n^*]^T$ while satisfying m inequality and n equality constraints as follows:

$$g_i(\bar{x}) \geq 0, i = 1, 2, ..., m \tag{10.11}$$

and

$$h_i(\bar{x}) \geq 0, i = 1, 2, ..., n \tag{10.12}$$

and the optimization vector is

$$f(\bar{x}) = [f_1(\bar{x}), f_2(\bar{x}), ..., f_M(\bar{x})]^T \tag{10.13}$$

While considering a maximization problem, a solution $\bar{x}_i$ is said to dominate $\bar{x}_j$ if the following conditions are satisfied.

$$\forall k \in 1, 2, ..., M, f_k(\bar{x}_i) \geq f_k(\bar{x}_j) \tag{10.14}$$

and

$$\exists k \in 1, 2, ..., M, s.t. f_k(\bar{x}_i) > f_k(\bar{x}_j) \tag{10.15}$$

The non-dominated set of solutions are those solutions, which are not dominated by any member of the set of solutions. This non-dominated set of solutions is called the Pareto optimal front.

10.4.1   NSGA-II

In the year 2002, a fast elitist Non-dominated Sorting Genetic Algorithm (NSGA-II) [15] was proposed by Deb, *et al.* It provides a mechanism for better sorting and incorporates elitism and diversity preservation mechanism to improve the performance of the NSGA algorithm [14]. It shows good performance in solving critical problems. The working principle of the NSGA-II algorithm for automatic clustering of gray scale images is discussed as follows:

10.4.2   POPULATION INITIALIZATION AND CHROMOSOME REPRESENTATION

Initially, a population P with N number of chromosomes has been produced by the normalized values (between 0 and 1) of randomly selected pixel intensity values of an input image. The length L of each chromosome has been taken as the square root of the highest intensity value of the input image.

10.4.3   CREATING CLUSTER CENTROIDS

A predefined threshold value $T_h = 0.5$ has been used to identify the active cluster centroids from each chromosome. In a particular generation t, for any chromosome, if the value of T_h is greater than the normalized intensity value of $P_{i,j}^t$, where $i = \{1, 2, ..., N\}$ and $j = \{1, 2, ..., L\}$, then that value is selected for the active cluster center of that chromosome.

10.4.4 GENETIC OPERATION

In this chapter, in order to implement classical NSGA-II for automatic clustering of gray scale images, three genetic operators, viz., tournament selection, conventional *crossover* and *mutation*, have been used. The elaborate description of the different genetic operations is available in [34]. The elitism operation has also been carried out to identify the next-generation population. The last generation provides different solutions, which are considered as the near-Pareto-optimal solutions.

10.4.5 FAST NON-DOMINATED SORTING

The new fitness values from both the objective functions have been evaluated after performing the conventional *crossover* and *mutation* operation. At the end of these operations, total $2N$ number of solutions has been generated. The first non-dominated sorting is then done on these $2N$ number of solutions to create the near Pareto-optimal front. A detailed explanation of the first non-dominated sorting procedure is available in [34].

10.4.6 CROWDING DISTANCE

Once the non-dominated sort has been completed, the population is sorted according to each objective function in an ascending order based on the computation of the crowding distance. The computational details of the crowding distance calculation is available in [34]. Thereafter, the first N number of solutions are considered from the front to prepare the next-generation population P^{t+1}.

10.4.7 BASIC STEPS OF CLASSICAL NSGA-II ALGORITHM FOR AUTOMATIC CLUSTERING OF GRAY SCALE IMAGES

The experimental steps of classical NSGA-II for automatic clustering of gray scale images are presented as follows:

1. A population P^t is created from the input gray scale image as described in Section 10.4.2.
2. The active cluster centroids of all the chromosomes are identified belonging to P^t as discussed in Section 10.4.3.
3. Both the fitness values $FV1$ by Equation (10.6) and $FV2$ by Equation (10.10) are evaluated simultaneously for all the chromosomes as explained in Section 10.3.
4. The tournament selection is performed and the population is updated by maintaining elitism as discussed in Section 10.4.4.
5. The conventional crossover operation is performed based on a predefined crossover probability to produce new offspring NP^t.
6. The conventional mutation is performed based on a predefined mutation probability over some randomly selected chromosomes belonging to $P^t \bigcup NP^t$.

7. After that, the fast non-dominated sorting is performed on $P^t \bigcup NP^t$ to produce near Pareto-optimal front as discussed in Section 10.4.5.
8. The crowding distance of all the elements from near Pareto-optimal front is conducted to identify the first N number of chromosomes to produce the next-generation population P^{t+1}. Both the fitness values, $FV1$, and $FV2$, along with their number of cluster centroids, corresponding to those solutions, are memorized as explained in Section 10.4.6.
9. Steps 4 to 8 are repeated until the stopping criteria is met.
10. Finally, the obtained output is reported.

10.5   PROPOSED TECHNIQUE

The proposed quantum-inspired multi-objective NSGA-II (QIMONSGA-II) algorithm for automatic clustering of gray scale images is elaborated below along with the flowchart, presented in Figure 10.4.

10.5.1   QUANTUM STATE POPULATION INITIALIZATION

In the proposed algorithm, the quantum state population Q is created to encode the original population P. Hence, Q consists of N number of strings and each string stores L number of quantum state information. The quantum state population is created by the realization of probabilistic representation of qubits [17]. A single *qubit* is defined with a pair of number, (α, β), as

$$\begin{bmatrix} \alpha \\ \beta \end{bmatrix} \tag{10.16}$$

where $|\alpha|^2 + |\beta|^2 = 1$. In a particular t^{th} generation, the quantum state population (Q^t) is defined as

$$Q^t = \{q_1^t, q_2^t, ..., q_N^t\} \tag{10.17}$$

where N is the size of the population and q_i^t, $i = \{1, 2, ..., N\}$ represents each individual chromosome. Therefore, each chromosome with length L belonging to Q^t can be defined as

$$q_{ij}^t = \begin{bmatrix} \alpha_{i1}^t & \alpha_{i2}^t & \cdots & \alpha_{iL}^t \\ \beta_{i1}^t & \beta_{i2}^t & \cdots & \beta_{iL}^t \end{bmatrix} \tag{10.18}$$

where $i = \{1, 2, ..., N\}$. Another way of representing Equation (10.18) is as follows:

$$q_{ij}^t = \begin{bmatrix} \cos\theta_{i1}^t & \cos\theta_{i2}^t & \cdots & \cos\theta_{iL}^t \\ \sin\theta_{i1}^t & \sin\theta_{i2}^t & \cdots & \sin\theta_{iL}^t \end{bmatrix} \tag{10.19}$$

where $i = \{1, 2, ..., N\}$ and L is the length of each string.

Initially, the quantum state population Q^t has been generated by random values of $\theta_{i,j}^t$, where $t = 0$. Each chromosome belonging to original population P^t has been encoded by some binary strings B_{iL}^t, where $i = \{1, 2, ..., N\}$ and L is the length of each string. The binary strings B_{iL}^t are then generated after observing the values of Q^t by the following equation.

$$B_{i,j}^t = \begin{cases} 1, & \text{if } \left|\beta_{i,j}^t\right|^2 > \left|\alpha_{i,j}^t\right|^2 \\ 0, & \text{Otherwise.} \end{cases} \tag{10.20}$$

where $i = \{1, 2, ..., N\}$ and $j = \{1, 2, ..., L\}$.

10.5.2 CREATING CLUSTER CENTROIDS IN QUANTUM-INSPIRED FRAME-WORK

The active cluster centroids are identified for each chromosome belonging to P^t by observing the values of B_{iL}^t, s.t. $\{\forall B_{iL}^t = 1\}$, $i = \{1, 2, ..., N\}$ and L is the length of each chromosome.

10.5.3 GENETIC OPERATORS IN QUANTUM-INSPIRED FRAMEWORK

Unlike NSGA-II, the proposed QIMONSGA-II uses quantum behaved selection, followed by quantum behaved crossover and quantum behaved mutation. The next generation is created from the *non-dominated* solutions among the parent and child populations by these *elitism* operations. The elaborate description of the different conventional genetic processes is available in [34]. Different solutions are identified from the near-Pareto-optimal strings of the last generation.

10.5.3.1 Quantum-Behaved Selection

The concept of *quantum rotation gate* operation has been used for selecting the most promising chromosome within a population. The quantum rotation gate operation has been applied on each individual belonging to Q^t in order to produce a new quantum state population NQ^t for the same generation t. The mathematical description of quantum rotation gate is given by

$$R(\delta\theta) = \begin{bmatrix} \cos(\Delta\theta) & -\sin(\Delta\theta) \\ \sin(\Delta\theta) & \cos(\Delta\theta) \end{bmatrix} \tag{10.21}$$

where $\Delta\theta$ is a very small rotation angle which has been taken randomly between $[-0.5, 0.5]$ for updating the value of each qubit in Q^t to produce NQ^t. It is depicted by

$$R(\delta\theta)\begin{bmatrix} \alpha \\ \beta \end{bmatrix} = R(\delta\theta)\begin{bmatrix} \cos\theta \\ \sin\theta \end{bmatrix} = \begin{bmatrix} \cos(\Delta\theta) & -\sin(\Delta\theta) \\ \sin(\Delta\theta) & \cos(\Delta\theta) \end{bmatrix}\begin{bmatrix} \cos\theta \\ \sin\theta \end{bmatrix}$$
$$= \begin{bmatrix} \cos(\theta + \delta\theta) \\ \sin(\theta + \delta\theta) \end{bmatrix} = \begin{bmatrix} \alpha' \\ \beta' \end{bmatrix} \tag{10.22}$$

Figure 10.1 depicts the effect of quantum rotation gate operation, which is responsible for creating new quantum states.

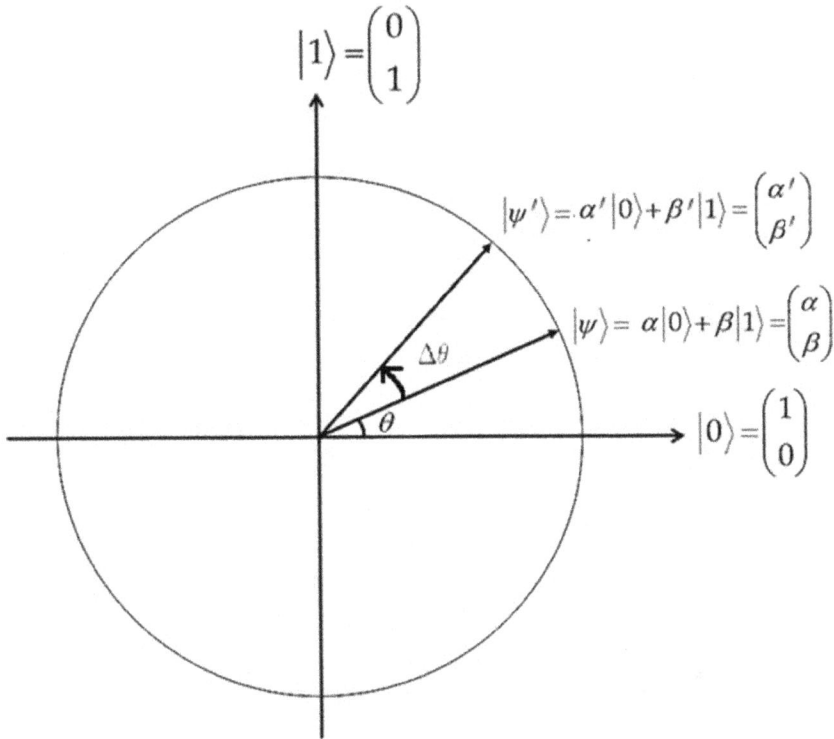

$$|1\rangle = \begin{pmatrix} 0 \\ 1 \end{pmatrix}$$

$$|\psi'\rangle = \alpha'|0\rangle + \beta'|1\rangle = \begin{pmatrix} \alpha' \\ \beta' \end{pmatrix}$$

$$|\psi\rangle = \alpha|0\rangle + \beta|1\rangle = \begin{pmatrix} \alpha \\ \beta \end{pmatrix}$$

$$|0\rangle = \begin{pmatrix} 1 \\ 0 \end{pmatrix}$$

$\Delta\theta$

θ

Figure 10.1: Quantum rotation gate operation.

Now, each individual from NQ^t produces new binary string NB^t_{iL}, where $i = \{1,2,...,N\}$ and L is the length of each chromosome. Then, each string belonging to NB^t is used to evaluate the new fitness from both the objective functions of each chromosome of P^t. Thereafter, the best solutions are identified between Q^t and NQ^t to update Q^t along with B^t. An extensive explanation of the working principle of quantum rotation gate operation is presented in [28][35].

10.5.3.2 Quantum-Behaved Crossover

The quantum-behaved *crossover* is performed on Q^t to create NQ^t. The crossover between any two individuals (say i and j, where $(i,j) \in N$ and $i \neq j$) from Q^t has been carried out depending upon a predefined crossover probability. Figure 10.2 depicts as to how any two individuals are responsible for creating new individuals.

Crossover Point

$$Q_i : \begin{bmatrix} \alpha_{i,1} & \alpha_{i,2} & \alpha_{i,3} & \alpha_{i,4} & \alpha_{i,5} & \alpha_{i,6} & \cdots & \alpha_{i,L} \\ \beta_{i,1} & \beta_{i,2} & \beta_{i,3} & \beta_{i,4} & \beta_{i,5} & \beta_{i,6} & \cdots & \beta_{i,L} \end{bmatrix}$$

$$Q_j : \begin{bmatrix} \alpha_{j,1} & \alpha_{j,2} & \alpha_{j,3} & \alpha_{j,4} & \alpha_{j,5} & \alpha_{j,6} & \cdots & \alpha_{j,L} \\ \beta_{j,1} & \beta_{j,2} & \beta_{j,3} & \beta_{j,4} & \beta_{j,5} & \beta_{j,6} & \cdots & \beta_{j,L} \end{bmatrix}$$

$$NQ_i : \begin{bmatrix} \alpha_{i,1} & \alpha_{i,2} & \alpha_{i,3} & \alpha_{j,4} & \alpha_{j,5} & \alpha_{j,6} & \cdots & \alpha_{j,L} \\ \beta_{i,1} & \beta_{i,2} & \beta_{i,3} & \beta_{j,4} & \beta_{j,5} & \beta_{j,6} & \cdots & \beta_{j,L} \end{bmatrix}$$

$$NQ_j : \begin{bmatrix} \alpha_{j,1} & \alpha_{j,2} & \alpha_{j,3} & \alpha_{i,4} & \alpha_{i,5} & \alpha_{i,6} & \cdots & \alpha_{i,L} \\ \beta_{j,1} & \beta_{j,2} & \beta_{j,3} & \beta_{i,4} & \beta_{i,5} & \beta_{i,6} & \cdots & \beta_{i,L} \end{bmatrix}$$

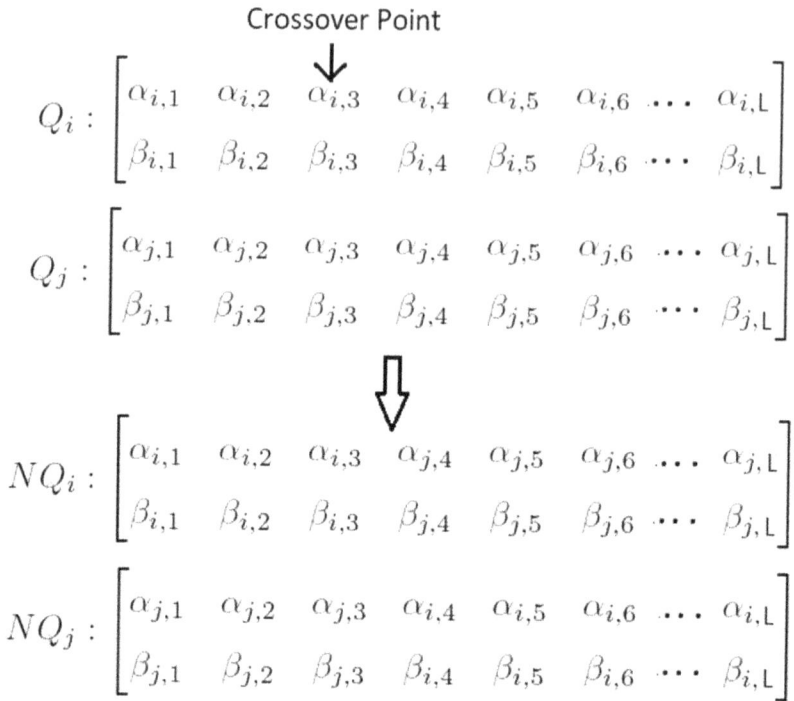

Figure 10.2: Quantum-behaved crossover.

10.5.3.3 Quantum-Behaved Mutation

The quantum-behaved *mutation* is implemented by utilizing the concept of Pauli-X gate. The Pauli-X gate operation has been performed over some qubits depending upon a predefined mutation probability, which reverses the probability amplitude values of a single *qubit* in Q^t. Mathematically *Pauli-X gate (PX)* can be defined as

$$PX = \begin{bmatrix} 0 & 1 \\ 1 & 0 \end{bmatrix} \tag{10.23}$$

The Pauli-X gate operation on a single qubit, responsible for reversing the probability amplitude values of that qubit, is demonstrated as follows:

$$PX \begin{bmatrix} \alpha \\ \beta \end{bmatrix} = \begin{bmatrix} 0 & 1 \\ 1 & 0 \end{bmatrix} \begin{bmatrix} \alpha \\ \beta \end{bmatrix} = \begin{bmatrix} \beta \\ \alpha \end{bmatrix} \tag{10.24}$$

Figure 10.3 depicts the effect of quantum behave mutation.

Mutation Point

$$Q_i : \begin{bmatrix} \alpha_{i,1} & \alpha_{i,2} & \alpha_{i,3} & \alpha_{i,4} & \alpha_{i,5} & \alpha_{i,6} & \cdots & \alpha_{i,L} \\ \beta_{i,1} & \beta_{i,2} & \beta_{i,3} & \beta_{i,4} & \beta_{i,5} & \beta_{i,6} & \cdots & \beta_{i,L} \end{bmatrix}$$

$$NQ_i : \begin{bmatrix} \alpha_{i,1} & \alpha_{i,2} & \alpha_{i,3} & \alpha_{i,4} & \beta_{i,5} & \alpha_{i,6} & \cdots & \alpha_{i,L} \\ \beta_{i,1} & \beta_{i,2} & \beta_{i,3} & \beta_{i,4} & \alpha_{i,5} & \beta_{i,6} & \cdots & \beta_{i,L} \end{bmatrix}$$

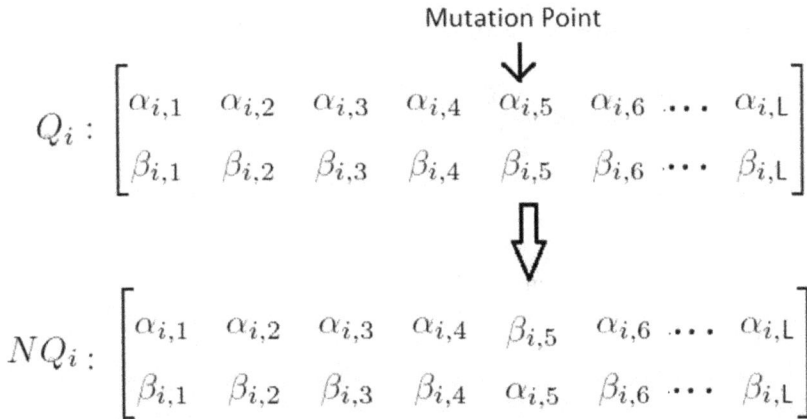

Figure 10.3: Quantum-behaved mutation.

10.5.4   FAST NON-DOMINATED SORTING IN QUANTUM-INSPIRED FRAMEWORK

The new fitness values from both the objective functions are evaluated after performing the quantum-behaved crossover and *mutation* operations. At the end of these operations, total $2N$ numbers of solutions are generated. Like classical NSGA-II algorithm, the first non-dominated sorting is done on these $2N$ numbers of solutions to create the near Pareto-optimal front. A detailed explanation of the fast non-dominated sorting procedure is available in [34].

10.5.5   CROWDING DISTANCE COMPUTATION IN QUANTUM-INSPIRED FRAMEWORK

Once the non-dominated sort has been completed, the crowding distance computation is required to sort the population according to each objective function in an ascending order. The computational details of the crowding distance calculation are similar to the classical NSGA-II algorithm, which is elaborately presented in [34]. Thereafter, the first N number of solutions have been considered from the front to prepare the next-generation population P^{t+1} and the corresponding qubits have been chosen for Q^{t+1}.

10.5.6   QIMONSGA-II ALGORITHM FOR AUTOMATIC CLUSTERING OF GRAY SCALE IMAGES

This section presents the proposed Quantum-Inspired Multi-Objective NSGA-II (QIMONSGA-II) algorithm for automatic clustering of gray scale images.

Input Parameters
    Maximum Generation:= $M_{ax}G$
    Population Size := N
    Crossover Probability := C_p
    Mutation Probability := μ_p

Output Parameters
    Optimum Cluster Number:= ON_C
    Optimum Fitness Value1 := $FV1$
    Optimum Fitness Value2 := $FV2$

1. $t \leftarrow 0$
2. Create original population P^t from the input gray scale image as described in Section 10.4.2
3. Create quantum state population Q^t to encode P^t as discussed in Section 10.5.1.
4. Identify active cluster centroids of all the chromosomes belonging to P^t by the guidance of Q^t as described in Section 10.5.2.
5. Evaluate both the fitness values, $FV1$ by Equation (10.6) and $FV2$ by Equation (10.10) simultaneously of all the chromosomes belonging to P^t as discussed in Section 10.3.
6. Perform quantum-behaved rotation gate operation on Q^t to create new NQ^t as elaborated in Section 10.5.3.1.
7. Again identify active cluster centroids of all the chromosomes belonging to P^t by the guidance of NQ^t.
8. Evaluate both the fitness values, $FV1$ and $FV2$, simultaneously of all the chromosomes belonging to P^t.
9. Identify N number of best solutions from $Q^t \cup NQ^t$ by performing fast non-dominated sorting followed by crowding distance calculation and update Q^t by those solutions along with their fitness values and number of cluster centroids.
10. Depending upon a predefined crossover probability C_p, perform quantum-behaved crossover operation on Q^t to produce offspring NQ^t as elaborated in Section 10.5.3.2.
11. Depending upon a predefined mutation probability μ_p, perform quantum-behaved mutation operation on some strings of Q^t as discussed in Section 10.5.3.3.
12. Perform fast non-dominated sorting followed by crowding distance calculation to generate the near Pareto-optimal front and thereafter consider the first N number of solutions from the front to prepare the next-generation population P^{t+1} with its corresponding quantum state population Q^{t+1}. Memorize the corresponding fitness values along with their number of cluster centroids as described in Sections 10.5.4 and 10.5.5 repeatedly.
13. $t \leftarrow t + 1$
14.   a. If $t < M_{ax}G$ then
      b. Repeat Steps from 6 to 14
15. Finally, report the obtained output.

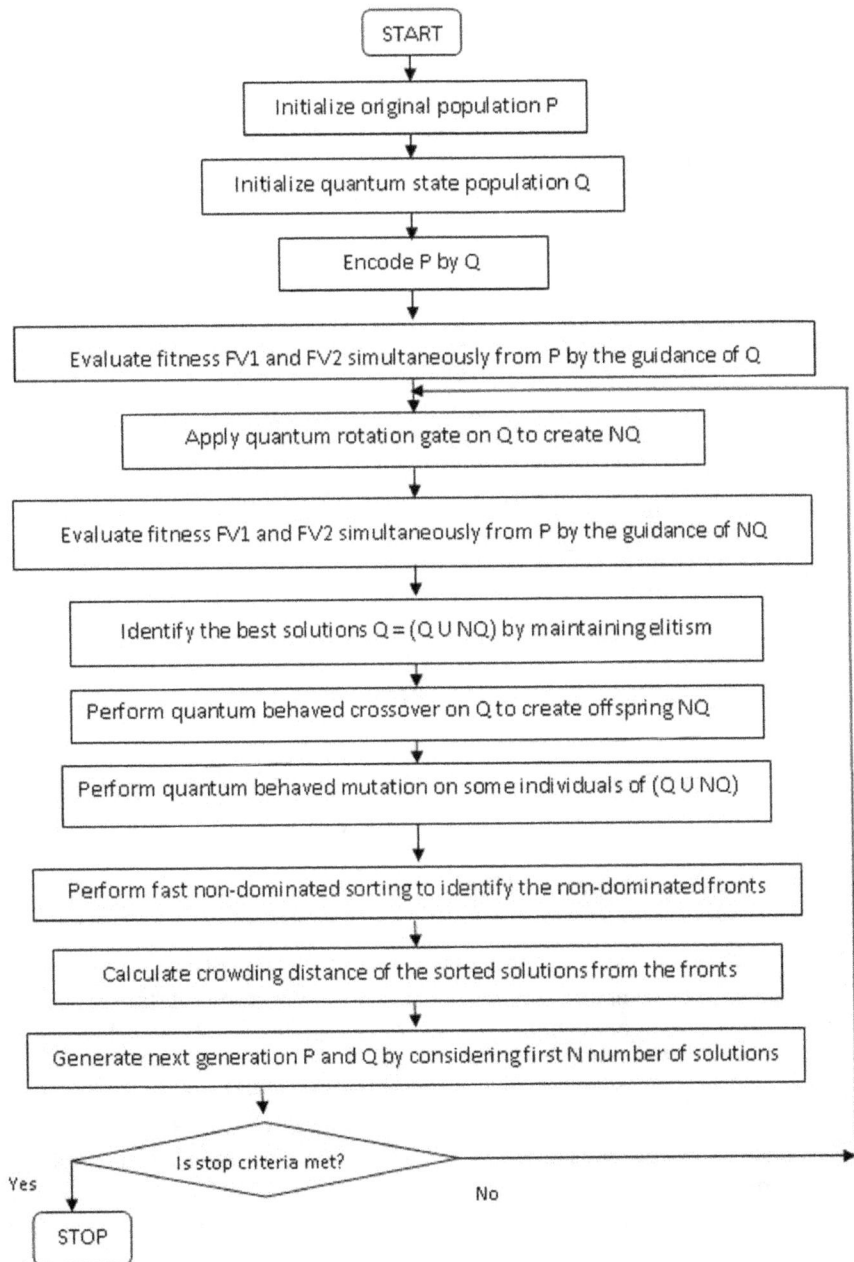

Figure 10.4: Flowchart of QIMONSGA-II algorithm for automatic clustering of gray scale images.

10.6   EXPERIMENTAL RESULTS AND ANALYSIS

This section presents an elaborate discussion of the experimental results of the proposed quantum-inspired multi-objective NSGA-II algorithm for automatic clustering of gray scale images along with its classical counter parts. All the experiments have been conducted on a personal computer with Intel(R) Core (TM) i5-8250U processor, 8.00GB RAM and 1.60GHz CPU speed. The executable programs have been developed in Windows 10 operating system and in Python language.

10.6.1   USED DATASET

In this paper, all the experiments have been conducted over six Berkeley [31] gray scale images of different sizes. Among these six test images, the first three images, viz., (a)#86000, (b)#92059 and (c)#89072 are of dimensions 321×481 and the last three images, viz., (d)#86016, (e)#87046 and (f)#94079 are of dimensions 481×321. All the test images are shown in Figure 10.5.

10.6.2   PARAMETER SETTINGS

The settings of the input parameters for both the algorithms, viz., QIMONSGA-II and NSGA-II for automatic clustering of gray scale images are presented in Table 10.1.

10.6.3   PERFORMANCE EVALUATION

In this paper, the well-known Minkowski score [36] has been evaluated for understanding the results of clustering. Mathematically, the Minkowski score (M_S) for a computed solution $(Comp)$ is defined as follows:

$$M_S = \frac{M^{True} - M^{Comp}}{M^{True}} \tag{10.25}$$

where M^{Comp} represents a binary matrix s.t. $M_{i,j}^{Comp} = 1$, iff i and j belong to the same cluster of a computed solution $(Comp)$ and M^{True} represents the corresponding matrix for the true solutions $True$. The better clustering results are achieved for the smaller values of this score.

Additionally, the Silhouette index (SIL) [37] has been used to measure the performance of clustering by the proposed algorithm. In the year 1987, Rousseeuw proposed the Silhouette index (SIL) [37]. It compares the pairwise difference of between and within cluster distances and is defined by

$$SIL = \frac{1}{K} \sum_{i=1}^{K} S(C_i) \tag{10.26}$$

(a)   (b)   (c)

(d)   (e)

(f)

Figure 10.5: Original test images [31]: (a)#86000, (b)#92059, (c)#89072, (d)#86016, (e)#87046, (f)#94079.

where the Silhouette width for the given cluster C_i is represented by $S(C_i)$. The value of $S(C_i)$ is computed as

$$S(C_i) = \frac{1}{N_i} \sum_{x \in C_i} \frac{b(x) - a(x)}{max(a(x), b(x))} \qquad (10.27)$$

Table 10.1

Input Parameters for QIMONSGA-II and NSGA-II [36]

Parameters	QIMONSGA-II	NSGA-II
Population Size :	50	50
Maximum Generation :	50	100
Crossover Probability :	0.8	0.8
Mutation Probability :	$\frac{1}{ChromosomeLength}$	$\frac{1}{ChromosomeLength}$
Small Rotation Angle :	[-0.1 to 0.1]	-

Table 10.2

Results of Mean Fitness Values from CSM [7] and DB [30]

Data Sets	CVI	QIMONSGA-II	NSGA-II
#86000	CSM	0.29673	0.33841
	DB	0.274e7	0.355e7
#92059	CSM	0.27115	0.36954
	DB	0.335e7	0.248e7
#89072	CSM	5.07643	6.32973
	DB	6.92591	7.19382
#86016	CSM	0.13794	0.26844
	DB	1.35473	1.90035
#87046	CSM	0.30133	0.31472
	DB	0.352e7	0.287e6
#94079	CSM	0.51692	0.39961
	DB	8.19665	8.94837

where N_i is the number of patterns belonging to C_i, $a(x)$ represents the within cluster mean distance by averaging the distance between x and the rest of the patterns from the same cluster, whereas, $b(x)$ represents the smallest mean distance of x to the patterns from another cluster. The *SIL* [37] index value generally lies between -1 to 1. The optimal result is achieved for the maximum value of *SIL* index.

10.6.4   EXPERIMENTAL RESULTS

Both the algorithms, viz., QIMONSGA-II and NSGA-II, have been executed for 30 runs. Table 10.2 presents the mean fitness values of both the objective functions, viz., CSM [7] and DB [30] corresponding to the best solutions which have been obtained from the final stage of non-dominated Pareto optimal front. The non-dominated Pareto optimal front obtained by QIMONSGA-II and NSGA-II are presented in Figures 10.6 and 10.7, respectively.

The result of standard deviation (σ), standard error (ε), and optimal computational time (τ in second) by CSM [7] and DB [30] are presented in Table 10.3. After

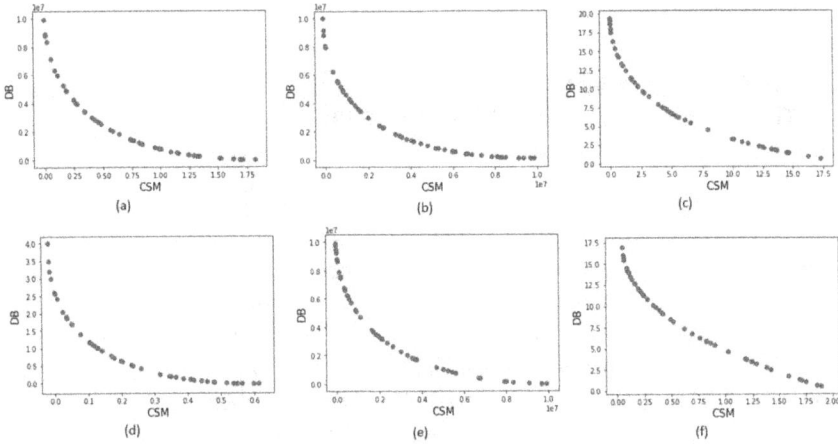

Figure 10.6: Non-dominating Pareto optimal front of test images [31] (a)#86000, (b)#92059, (c)#89072, (d)#86016, (e)#87046, (f)#94079 by QIMONSGA-II.

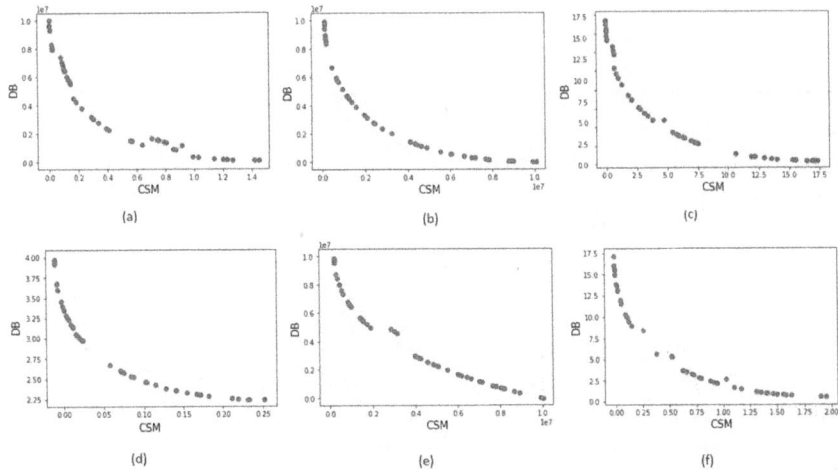

Figure 10.7: Non-dominating Pareto optimal front of test images [31] (a)#86000, (b)#92059, (c)#89072, (d)#86016, (e)#87046, (f)#94079 by NSGA-II.

analyzing the results of Table 10.3, it is established that the proposed algorithm performs better than its classical counterpart as most of the results are in favor of the proposed algorithm.

While performing the unpaired t-$test$ [38] between QIMONSGA-II and NSGA-II [15], it is seen that eight results are "extremely significant," two are "very significant" and one is "significant" but one result is identified as "not significant" result. The result of unpaired t-$test$ [38] is presented in Table 10.4.

Table 10.3

Results of Standard Deviation (σ), Standard Error (ε) and Optimal computational Time (τ in second) by CSM [7] and DB [30]

Data Sets	CVI	QIMONSGA-II			NSGA-II		
		σ	ε	τ	σ	ε	τ
#86000	CSM	**0.02682**	**0.00489**	**63**	0.03783	0.00504	97
	DB	**0.01775**	**0.00324**	**85**	0.02764	0.00573	118
#92059	CSM	**0.00784**	**0.00143**	**54**	0.05941	0.01084	72
	DB	**0.03965**	**0.00729**	77	0.08451	0.01543	**65**
#89072	CSM	0.07834	0.01431	**61**	**0.01919**	**0.00350**	89
	DB	**0.23968**	**0.04375**	67	0.38936	0.07108	94
#86016	CSM	**0.16543**	**0.03020**	**45**	0.25828	0.04715	94
	DB	**0.08164**	**0.01492**	**59**	0.09739	0.01778	78
#87046	CSM	0.49472	0.09032	**56**	**0.20398**	**0.03726**	**56**
	DB	**0.03657**	**0.00667**	**83**	0.05826	0.01064	126
#94079	CSM	**0.21943**	**0.04006**	**38**	0.22559	0.04186	61
	DB	0.05828	0.01064	**52**	**0.04694**	**0.00857**	87

Table 10.4

Results of Unpaired *t-test* [38] between QIMONSGA-II and NSGA-II for CSM [7] and DB [30]

Data Sets	CVI	QIMONSGA-II vs. NSGA-II	
		$P-Value$	Significance Level
#86000	CSM	<0.0001	Extremely Significant
	DB	<0.0001	Extremely Significant
#92059	CSM	<0.0001	Extremely Significant
	DB	<0.0001	Extremely Significant
#89072	CSM	<0.0001	Extremely Significant
	DB	0.0022	Very Significant
#86016	CSM	0.0233	Significant
	DB	<0.0001	Extremely Significant
#87046	CSM	0.8915	Not Significant
	DB	<0.0001	Extremely Significant
#94079	CSM	0.0457	Significant
	DB	<0.0001	Extremely Significant

Table 10.5 presents the obtained values of M_S [36] and *SIL* [37], which have been used to compare the performance of the proposed algorithm with its classical counterpart. While considering the score value of M_S [36] and *SIL*, it is found that all the test images excluding #87046 have scored better M_S [36] score values and *SIL*

Table 10.5

Results of Performance Evaluation by M_S [36] and *SIL* [37]

Data Sets	Performance Metrics	QIMONSGA-II	NSGA-II
#86000	M_S	**0.50127**	0.58982
	SIL	**0.65294**	0.52369
#92059	M_S	**0.39421**	0.43024
	SIL	**0.58392**	0.44993
#89072	M_S	**0.36375**	0.36333
	SIL	**0.70025**	0.65047
#86016	M_S	**0.53285**	0.59346
	SIL	**0.68327**	0.63284
#87046	M_S	0.47293	**0.49211**
	SIL	0.45982	**0.49328**
#94079	M_S	**0.32691**	0.43973
	SIL	**0.68485**	0.47226

[37] values. This proves the supremacy of the proposed algorithm over its classical counterpart.

Finally, the clustered images obtained from the proposed algorithm and its classical counterparts are presented in Figures 10.8 and 10.9, respectively. The corresponding threshold values obtained by QIMONSGA-II, used for creating the clustered images, are presented in Table 10.6.

10.7 DISCUSSIONS AND CONCLUSION

In this paper, a Quantum-Inspired Multi-Objective NSGA-II (QIMONSGA-II) algorithm has been presented for automatic clustering of gray scale images. In this work, the proposed algorithm has been compared with its classical counterpart. All the experiments have been performed over six Berkeley gray scale images. The superiority of the proposed algorithm over its classical counterpart has been justified by some metrics, viz., computational time, mean fitness values obtained from both the cluster validity indices, viz., CSM [7] and DB [30], standard deviation, and standard error. Moreover, a statistical superiority tests, viz., unpaired *t-test* has been performed to prove the efficacy of the proposed algorithm. Finally, the superiority of the proposed algorithm has been proved with the help of Minkowski score [36] and Silhouette index (*SIL*) [37]. Though the proposed algorithm has been applied only on gray scale Berkeley images, there still remains a scope of research for using them on true color images. The authors are currently engaged in this direction.

(a)   (b)   (c)

(d)   (e)

(f)

Figure 10.8: Clustered images corresponding to test images [31]: (a)#86000, (b)#92059, (c)#89072, (d)#86016, (e)#87046, (f)#94079, by QIMONSGA-II.

(a)   (b)   (c)

(d)   (e)

(f)

Figure 10.9: Clustered images corresponding to test images [31]: (a)#86000, (b)#92059, (c)#89072, (d)#86016, (e)#87046, (f)#94079, by NSGA-II.

Table 10.6

Results of Number of Clusters and Threshold Values

Data Sets	Number of Clusters	Threshold Values
#86000	3	[55, 102, 180]
#92059	4	[46, 89, 119, 223]
#89072	3	[56, 120, 193]
#86016	4	[82, 140, 176, 209]
#87046	4	[75, 112, 127, 239]
#94079	5	[57, 103, 141, 168, 217]

REFERENCES

1. A. K. Jain and R. C. Dubes. Algorithms for Clustering Data. Prentice-Hall, Inc., USA, 1988.

2. A. K. Jain, M. N. Murty, and P. J. Flynn. Data clustering: A review. ACM Computing Surveys, 31(3):264–323, 1999.

3. Ashwini Gulhane, Prashant Paikrao, and D. Chaudhari. A review of image data clustering techniques. International Journal of Soft Computing and Engineering (IJSCE), 2(1):212–215, 2011.

4. J. C. Platt, M. Czerwinski, and B.A. Field. Phototoc: Automatic clustering for browsing personal photographs. In Fourth International Conference on Information, Communications and Signal Processing, 2003 and the Fourth Pacific Rim Conference on Multimedia. Proceedings of the 2003 Joint, volume 1, pages 6–10, 2003.

5. J.H. Chen, Y.C. Chang, and W.L. Hung. A robust automatic clustering algorithm for probability density functions with application to categorizing color images. Communications in Statistics – Simulation and Computation, 47(7):2152–2168, 2018.

6. T. Geraud, P. Strub, and J. Darbon. Color image segmentation based on automatic morphological clustering. In Proceedings 2001 International Conference on Image Processing (Cat. No.01CH37205), volume 3, pages 70–73, 2001.

7. T. Lei, P. Liu, X. Jia, X. Zhang, H. Meng, and A.K. Nandi. Automatic fuzzy clustering framework for image segmentation. IEEE Transactions on Fuzzy Systems, 28(9):2078–2092, 2020.

8. S. Bandyopadhyay and U. Maulik. Genetic clustering for automatic evolution of clusters and application to image classification. Pattern Recognition, 35(6):1197–1208, 2002.

9. S. Das, A. Abraham, and A. Konar. Automatic clustering using an improved differential evolution algorithm. IEEE Transactions on Systems, Man, and Cybernetics – Part A: Systems and Humans, 38(1):218–237, 2008.

10. A.E. Ezugwu. Nature-inspired metaheuristic techniques for automatic clustering: A survey and performance study. SN Applied Sciences, 2, 2020.

11. A. Jose-Garca and W. Gomez-Flores. Automatic clustering using nature-inspired metaheuristics: A survey. Applied Soft Computing, 41:192–213, 2016.

12. S. Saha and S. Bandyopadhyay. A generalized automatic clustering algorithm in a multiobjective framework. Applied Soft Computing, 13:89–108, 2013.

13. K. Suresh, D. Kundu, S. Ghosh, S. Das, and A. Abraham. Data clustering using multiobjective differential evolution algorithms. Fundamenta Informaticae, 97:381–403, 2009.

14. N. Srinivas and K. Deb. Muiltiobjective optimization using nondominated sorting in genetic algorithms. Evolutionary Computation, 2(3):221–248, 1994.
15. K. Deb, A. Pratap, S. Agarwal, and T. Meyarivan. A fast and elitist multiobjective genetic algorithm: Nsgaii. IEEE Transactions on Evolutionary Computation, 6:182–197, 2002.
16. D.W. Corne, N.R. Jerram, J.D. Knowles, and M.J. Oates. PESA-II: Region-based selection in evolutionary multiobjective optimization. In Proceedings of the 3rd Annual Conference on Genetic and Evolutionary Computation, GECCO01, pages 283–290, San Francisco, CA, USA, 2001. Morgan Kaufmann Publishers Inc.
17. E. Zitzler and L. Thiele. Multiobjective evolutionary algorithms: A comparative case study and the strength pareto approach. IEEE Transactions on Evolutionary Computation, vol. 3, no. 4, 257–271, 1999.
18. M. Kim, T. Hiroyasu, M. Miki, and S. Watanabe. SPEA2+: Improving the performance of the strength pareto evolutionary algorithm 2. In Xin Yao, Edmund K. Burke, Jose A. Lozano, Jim Smith, Juan Julian Merelo-Guervos, John A. Bullinaria, Jonathan E. Rowe, Peter Tino, Ata Kaban, and Hans-Paul Schwefel, editors, Parallel Problem Solving from Nature – PPSN VIII, pages 742–751, Springer, Berlin, Heidelberg, 2004.
19. U. Maulik, S. Bandyopadhyay, and A. Mukhopadhyay. Multiobjective genetic algorithms for clustering: Applications in data mining and bioinformatics. Springer Science & Business Media, 2011.
20. Aimin Zhou, Bo-Yang Qu, Hui Li, Shi-Zheng Zhao, Ponnuthurai Nagaratnam Suganthan, and Qingfu Zhang. Multiobjective evolutionary algorithms: A survey of the state of the art. Swarm and Evolutionary Computation, 1(1):32–49, 2011.
21. T. Hey. Quantum computing: An introduction. Computing & Control Engineering Journal, 10:105–112, June 1999.
22. K.H. Han and J.H. Kim. Quantum-inspired evolutionary algorithm for a class of combinatorial optimization. IEEE Transactions on Evolutionary Computation, 6(6):580–593, 2002.
23. T. Gandhi, Nitin, and T. Alam. Quantum genetic algorithm with rotation angle refinement for dependent task scheduling on distributed systems. In 2017 Tenth International Conference on Contemporary Computing (IC3), pages 1–5. IEEE, Aug 2017.
24. H.P. Chiang, Y.H. Chou, C.H. Chiu, S.Y. Kuo, and Y.M. Huang. A quantum-inspired tabu search algorithm for solving combinatorial optimization problems. Soft Computing, 18:1771–1781, 2013.
25. M. Ross and H. Oscar. A review of quantum-inspired metaheuristics: Going from classical computers to real quantum computers. IEEE Access, 8:814–838, 2019.
26. C. Wojciech and K. Joanna. Quantum-inspired evolutionary approach for the quadratic assignment problem. Entropy, 20(10):781, Oct 2018.
27. S. Dey, S. Bhattacharyya, and U. Maulik. Quantum inspired automatic clustering for multilevel image thresholding. In 2014 International Conference on Computational Intelligence and Communication Networks, pages 247–251, 2014.
28. A. Dey, S. Dey, S. Bhattacharyya, J. Platos, and V. Snasel. Novel quantum inspired approaches for automatic clustering of gray level images using particle swarm optimization, spider monkey optimization and ageist spider monkey optimization algorithms. Applied Soft Computing, 88(106040), 2020.
29. C.-H. Chou, M.-C. Su, and E. Lai. A new cluster validity measure and its application to image compression. Pattern Analysis and Applications, 7(2):205–220, Jul 2004.
30. D.L. Davies and D.W. Bouldin. A cluster separation measure. IEEE Transactions on Pattern Analysis and Machine Intelligence, 1:224–227, February 1979.
31. Berkley images. Accessed on 15/12/2020.

32. R. Blatt, H. Haiffner, C.F. Roos, C. Becher, and F. Schmidt-Kaler. Course 5 – quantum information processing in ion traps. In Daniel EstOEeve, Jean-Michel Raimond, and Jean Dalibard, editors, Quantum Entanglement and Information Processing, volume 79 of Les Houches, pages 223–260. Elsevier, 2004.

33. S. Bandyopadhyay, S. Saha, U. Maulik, and K. Deb. A simulated annealing-based multi-objective optimization algorithm: AMOSA. IEEE Transactions on Evolutionary Computation, 12(3):269–283, 2008.

34. K. Deb. Multiobjective Optimization Using Evolutionary Algorithms. Wiley, New York, 2001.

35. A. Dey, S. Dey, S. Bhattacharyya, J. Platos, and V. Snasel. Quantum inspired meta-heuristic approaches for automatic clustering of color images. International Journal of Intelligent Systems, 2021.

36. A. Mukhopadhyay, S. Bandyopadhyay, and U. Maulik. Clustering using multi-objective genetic algorithm and its application to image segmentation. In 2006 IEEE International Conference on Systems, Man and Cybernetics, vol. 3, pp. 2678–2683, 2006.

37. P.J. Rousseeuw. Silhouettes: A graphical aid to the interpretation and validation of cluster analysis. Journal of Computational and Applied Mathematics, 20:53–65, 1987.

38. B. Flury. A First Course in Multivariate Statistics. Springer Texts in Statistics.

11 Conclusion

A metaheuristic is a heuristic (partial search) algorithm that is more or less an efficient optimization algorithm to real-world problems. Hybrid metaheuristics refer to a proper and judicious combination of several other metaheuristics and machine learning algorithms. The hybrid metaheuristics have been found to be more robust and failsafe owing to the complementary character of the individual metaheuristics in the resultant combination. This is primarily due to the fact that the vision of hybridization is to combine different metaheuristics such that each of the combination supplements the other in order to achieve the desired performance.

Quantum computer, as the name suggests, principally works on several quantum physical features. These could be used as an immense alternative to today's apposite computers since they possess faster processing capability (even exponentially) than classical computers. A number of researchers coupled the underlying principles of quantum computing into various metaheuristic structures to introduce different quantum-inspired algorithmic approaches [1]–[5]. The evolution of the quantum computing paradigm has led to the evolution of time efficient and robust hybrid metaheuristics by means of conjoining the principles of quantum mechanics with the conventional metaheuristics, thereby enhancing the real-time performance of the hybrid metaheuristics.

This volume is a novel effort to bring together the recent advances and trends in designing new and novel quantum-inspired metaheuristics to solve real-life problems in various branches of science and engineering. This volume introduces the principles of quantum mechanics to evolve hybrid metaheuristics-based optimization techniques useful for real-world engineering and scientific problems. Starting from the introductory chapter, which presents an outline of the basic theory and concepts pertaining to quantum-inspired metaheuristics, the chapter also throws light on several types of quantum-inspired metaheuristics in details. It also comes up with a bird's eye view on the different bi-level/multi-level quantum system-based optimization techniques. In addition to that, several entanglement-induced optimization techniques and W-state encoding of optimization methods have also been discussed. The applications related to the theme of the topic have been provided that would also certainly be bring up to date the readers.

With the development of machine learning theory and the accumulation of practical experience of using various algorithms, it became clear that there is no ideal classification method that would be better than all others for all sizes of the training sample, for any percentage of noise in data, for any complexity of the boundaries of dividing objects into classes, etc. Therefore, at present, ensemble classification methods that combine many different classifiers trained on different data samples. One of the most accurate and fast parallelization methods available today is bagging, which turns out to be useful in the case of heterogeneous classifiers and instability, when small changes in the initial sample lead to significant changes in the classification.

DOI: 10.1201/9781003283294-11

Of late, several quantum-inspired methods for collective decision-making based on metaheuristic quantum algorithms have been proposed. These algorithms are found to increase the speed of combining decisions of basic classifiers.

As per definition, optimization is the technique of finding an alternative with the most cost-effective or highest achievable performance under the given constraints, by maximizing desired factors and minimizing undesired ones. Since many years, lot of efforts have been incorporated to resolve the optimization problems, specifically the NP-hard and multi-objective problems. It becomes slight easy after the invention of quantum computer, which is very costly and sensitive. IBM makes it available freely for the common people by introducing IBM Q Experience online platform [6]. The IPB Q circuit composer and QISkit can be used to resolve optimization problems using IBM Q.

Minimization of power loss in a DN is one of the challenging areas of research for the distribution utilities. In recent times, power losses are reduced by implementing DGs into distribution network. However, majority of research has been done on this important optimization problem with CP load model. Majority of consumers at load center uses VDLMs such as CZ, CC, IL, RL, and CL, whereas CP load model is independent of voltage. If the optimal placement and capacity of DG with CP load model are used on practical distribution system, it induces high power losses and poor voltage regulation in the system. In this study, an investigation has been performed to reduce the losses in the distribution system with DG for different VDLMs. Optimal location and capacity of DG is a difficult nondifferentiable, non-linear, complex combinatorial optimization problem. A Multipartite Adaptive Quantum-inspired Evolutionary Algorithm is proposed for optimal location and sizing of DG. MAQiEA uses probabilistic approach with Q-bits, and is an updated version of AQiEA that has introduced two Q-bits per solution vector and entanglement inspired adaptive crossover operator. MAQiEA has introduced a Multipartite Adaptive Crossover operator as a variation operator for better convergence.

A Quantum-Inspired Manta Ray Foraging Optimization algorithm can be evolved for automatic clustering of color images. A novel quantum rotation gate and Pauli-X gate strategy is used to achieve the exploration and exploitation. Several tests conducted among the competitive algorithms establish the effectiveness of the proposed algorithm.

A Quantum Genetic Algorithm (QGA) can be evolved to perform an automatic feature Cancer Modelling and Simulation selection in order to select only those features that has a strong influence on a Support Vector Machine-based (SVM) classifier. The QGA performs a search over the space formed by the feature set in order to find an optimal combination of features and at the same time, keeping or decreasing the loss rate in the training stage. The test image database is balanced in terms of the positive and negative stenosis cases. After the feature selection process ends, a subset of the features is able to keep the classification rate in terms of the accuracy metric and the Jaccard index, compared with the original set with all the features. In addition, the reduction of features has effect on the time required to perform an exhaustive feature extraction of new angiograms. The proposed method can be applied in clinical practice to assists cardiologists in the evaluation and finding of possible stenosis cases in X-ray coronary angiograms.

A Hybrid Quantum-Convolutional Neural Network method for atherosclerosis detection in XCA images can be envisaged. It includes a Quantum Convolutional Layer used as a preprocessing to improve the atherosclerosis detection performance of typical CNNs. Numerical experiments, based on two distinct CNN architectures (DenseNet-based and VGG-based architectures) and trying on two different training algorithms: Stochastic Gradient Descent and Stochastic Gradient Descent with Momentum, have demonstrated that using a Quantum Convolutional Layer on a limited XCA dataset performs efficiently for atherosclerosis detection. The introduced hybrid methodology improved the five evaluation metrics concerning the classical CNN architectures trained with the normalized-raw XCA images.

Both qubit- and qutrit-based Elephant Herd Optimization algorithms are efficient for automatic clustering of hyperspectral images. A modified rotation gate operation enhances the diversity of the population. The exploration and exploitation capabilities of the classical EHO algorithm are enhanced, with a faster convergence capability. As automatic cluster detection is a tedious task in HSI processing, these algorithms can be highly beneficial in real-life scenarios.

A novel Multiobjective Quantum-inspired Salp Swarm Algorithm (MQSSA) with Delta potential-well model presentation, a better alternative than the binary presentation for the multiobjective optimization problems, can be envisaged. The proposed approach is evaluated on several multiobjective optimization benchmark problems having convex-shaped and concave-shaped Pareto-optimal front. It shows a favorable outcome, with better performance than other well-regarded algorithms in the multiobjective domain.

Finally, a Quantum-Inspired Multi-Objective NSGA-II (QIMONSGA-II) algorithm has been presented for automatic clustering of gray scale images. The proposed algorithm has been compared with its classical counterpart. All the experiments have been performed over six Berkeley gray scale images. The superiority of the proposed algorithm over its classical counterpart has been justified by some metrics, viz., computational time, mean fitness values obtained from two cluster validity indices, viz., CSM [7] and DB [8], standard deviation, and standard error. This volume is expected to come to the benefit of senior researchers, practitioners, and aspiring researchers, who intend to work on quantum metaheuristic foundations and applications.

REFERENCES

1. Wang, L. & Niu, Q. & Fei, M. R. (2008). A novel quantum ant colony optimization algorithm and its application to fault diagnosis. Transactions of the Institute of Measurement and Control, 30(3–4), 313–329.
2. Dey, S. & Bhattacharyya, S. & Maulik, U. (2013). Quantum-inspired metaheuristic algorithms for multi-level thresholding for true colour images. In Proceeding of 2013 Annual IEEE India Conference (INDICON).
3. Dey, S. & Bhattacharyya, S. & Maulik, U. (2017). Efficient quantum-inspired metaheuristics for multi-level true colour image thresholding. Applied Soft Computing, 56, 472–513.

4. Dey, S. & Bhattacharyya, S. & Maulik, U. (2016). New quantum-inspired metaheuristic techniques for multi-level colour image thresholding. Applied Soft Computing, 46, 677–702.

5. https://qiskit.org/

6. Mishra, N. & Bisarya, A. et. al. (2020). Breast Cancer Detection Using Quantum Convolutional Neural Networks: A Demonstration on a Quantum Computer. medRxiv.

7. Chou, C.-H. & Su, M.-C. & Lai, E. (2004). A new cluster validity measure and its application to image compression. Pattern Analysis and Applications, 7(2), 205–220.

8. Davies, D. L. & Bouldin, D. W. (1979). A cluster separation measure. IEEE Transactions on Pattern Analysis and Machine Intelligence, 1, 224–227.

A Automatic Feature Selection for Coronary Stenosis Detection in X-Ray Angiograms Using Quantum Genetic Algorithm

A.1 MATLAB CODE TO EXTRACT VESSEL SEGMENTS

```
function segments = findsegments(m, endpoints)
%Find Segments Function
% Find the segments in a logical matrix using the
% information present in the neighborhood of a
% center element.
%--------------------------------------------------
%Artifact:  findsegments.m
%Version:   1.0
%Date:      15/03/2020 12:20:00
%Author:    Miguel Angel Gil Rios
%Email:     angel.grios@gmail.com
%--------------------------------------------------
%Usage:
%   segments = findsegments(m, endpoints)
%
%Inputs:
%           m:  The logical matrix
%   endpoints:  A Nx2 matrix containing the N
%               endpoints positions which this
%               function will try to follow a
%               connection path between them.
%--------------------------------------------------
```

DOI: 10.1201/9781003283294-A

```
%Outputs:
%     segments:   A Nx4 matrix containing the
%                 segment path located by the row
%                 and column index. Third column
%                 represents the  corresponding
%                 pixel slope. Fourth column
%                 represents the resultant angle
%                 in order to search structure
%                 borders.
%-------------------------------------------------
segments = {};
k = 0;

% Check if at least two end-points exists in the
% endpoints matrix:
if size(endpoints, 1) < 2 || size(endpoints, 2) ~= 2
return;
end
bp = zeros(0);
%Find segments while there exists endpoints:
while size(endpoints, 1) > 1

%Take the first endpoint:
start_point = endpoints(1, :);

%Find the segment:
[current_segment, m, end_point, bp] = ...
findsegment(m, start_point(1), start_point(2), bp);

%Check if a segment was formed:
if end_point ~= start_point

%Store the segment:
k = k + 1;
segments{k}.points = current_segment;

%Check if end_point exists in endpoints matrix:
pos = frindex(m, end_point);
if pos > 0
%Remove end-point:
endpoints(pos, :) = [];
else
%Replace end point:
endpoints(1, :) = end_point;
```

```
end
else
%Remove end-point:
endpoints(1, :) = [];
end
end


%After segments were computed, each pixel angle
%must be calculated.
%For each segment
for k = 1 : size(segments, 2)
v1 = segments{k}.points(1, :);
v2 = segments{k}.points(2, :);
[slope, angle] = computeSlope(v1, v2);
segments{k}.points(1, 3) = slope;
segments{k}.points(1, 4)= angle;
for kp = 2 : size(segments{k}.points, 1) - 1
v1 = segments{k}.points(kp - 1, :);
v2 = segments{k}.points(kp, :);
v3 = segments{k}.points(kp + 1, :);
[slope, angle] = computeSlope(v1, v2, v3);
segments{k}.points(kp, 3) = slope;
segments{k}.points(kp, 4) = angle;
end
i2 = size(segments{k}.points, 1);
i1 = i2 - 1;
v1 = segments{k}.points(i1, :);
v2 = segments{k}.points(i2, :);
[slope, angle] = computeSlope(v1, v2);
segments{k}.points(i2, 3) = slope;
segments{k}.points(i2, 4) = angle;
end
end %End Function

function [slope, angle] = computeSlope(v1, v2, v3)
%
%~~~~~~~~~~~~~~~~~~~~~~~~~~~~~~~~~~~~~~~~~~~~~~~~~~~~~~~~~~~~~~~~
%
%                    S1 = 0,    S2 = 0
%    m=0      m=0                                  angle=0 Degrees
%[v1]----[v2]----[v3]
%~~~~~~~~~~~~~~~~~~~~~~~~~~~~~~~~~~~~~~~~~~~~~~~~~~~~~~~~~~~~~~~~
%
%                    S1 = 0,    S2 = 1
%                                      m=0
%             [v3]                [v1]----[v2]
```

```
%                  /                          /            angle=45 Degrees
%              / m=1                      / m=1
% [v1]----[v2]                           [v3]
%      m=0
%~~~~~~~~~~~~~~~~~~~~~~~~~~~~~~~~~~~~~~~~~~~~~~~~~~~~~~~~~~~~~~~~~~~~
%                  S1 = 0,    S2 = Inf
%                                      m=0
%          [v3]                   [v1]----[v2]
%           | m=Inf                   | m=Inf  angle=90 Degrees
%           |                         |
%[v1]----[v2]                        [v3]
%      m=0
%~~~~~~~~~~~~~~~~~~~~~~~~~~~~~~~~~~~~~~~~~~~~~~~~~~~~~~~~~~~~~~~~~~~~
%                  S1 = 0,    S2 = -1
%                                      m=0
%          [v3]                   [v1]----[v2]
%           \ m=-1                   \ m=-1       angle=135 Degrees
%            \                        \
%[v1]----[v2]                        [v3]
%      m=0
%~~~~~~~~~~~~~~~~~~~~~~~~~~~~~~~~~~~~~~~~~~~~~~~~~~~~~~~~~~~~~~~~~~~~
%                  S1 = Inf,   S2 = Inf
%    [v1]
%     | m=Inf
%     |                                          angle=0 Degrees
%    [v2]
%     | m=Inf
%     |
%    [v3]
%~~~~~~~~~~~~~~~~~~~~~~~~~~~~~~~~~~~~~~~~~~~~~~~~~~~~~~~~~~~~~~~~~~~~
%                  S1 = Inf,   S2 = 0
%    [v1]                           [v1]
%     | m=Inf                        |m=Inf
%     |                              |           angle=90 Degrees
%    [v2]----[v3]                   [v3]----[v2]
%          m=0                           m=0
%~~~~~~~~~~~~~~~~~~~~~~~~~~~~~~~~~~~~~~~~~~~~~~~~~~~~~~~~~~~~~~~~~~~~
%                  S1 = Inf,   S2 = 1
%      [v1] [v3]                    [v1]
%   m=Inf |  /                       | m=Inf
%        | / m=1                     |           angle=45 Degrees
%      [v2]                         [v2]
%                                    /
%                                   / m=1
```

```
%                                  [v3]
%~~~~~~~~~~~~~~~~~~~~~~~~~~~~~~~~~~~~~~~~~~~~~~~~~~~~~~~~~~~~~~~~~~
%
%                  S1 = Inf,    S2 = -1
%         [v3] [v1]                        [v1]
%       m=-1 \ |m=Inf                      | m=Inf
%            \|                            |              angle=135 Degrees
%           [v2]                          [v2]
%                                            \
%                                             \ m=-1
%                                             [v3]
%~~~~~~~~~~~~~~~~~~~~~~~~~~~~~~~~~~~~~~~~~~~~~~~~~~~~~~~~~~~~~~~~~~
%         [v1]
%            \ m=-1
%             \
%           [v2]
%               \ m=-1
%                \
%               [v3]
%~~~~~~~~~~~~~~~~~~~~~~~~~~~~~~~~~~~~~~~~~~~~~~~~~~~~~~~~~~~~~~~~~~
%               [v1]
%               / m=1
%              /
%           [v2]
%            / m=1
%           /
%         [v3]
%

s1 = getSlope(v1, v2);
if ~exist('v3')
slope = s1;
switch slope
case 0
angle = 90;
case 1
angle = 135;
case -1
angle = 45;
case Inf
angle = 0;
otherwise
angle = 99;
end
else
```

```
s2 = getSlope(v2, v3);
if s1 == 0 && s2 == 0
slope = 0;
angle = 90;
elseif (s1 == 0 && s2 == 1) || (s2 == 0 && s1 == 1)
slope = 0;
angle = 90;
elseif (s1 == 0 && s2 == Inf) || (s2 == 0 && s1 == Inf)
slope = 1;
angle = 135;
elseif (s1 == 0 && s2 == -1) || (s2 == 0 && s1 == -1)
slope = 0;
angle = 90;
elseif s1 == Inf && s2 == Inf
slope = Inf;
angle = 0;
elseif (s1 == Inf && s2 == 0) || (s2 == Inf && s1 == 0)
slope = 1;
angle = 45;
elseif (s1 == Inf && s2 == 1) || (s2 == Inf && s1 == 1)
slope = Inf;
angle = 0;
elseif (s1 == -1 && s2 == Inf) || (s2 == -1 && s1 == Inf)
slope = Inf;
angle = 0;
elseif s1 == -1 && s2 == -1
slope = -1;
angle = 45;
elseif s1 == 1 && s2 == 1
slope = 1;
angle = 135;
else
slope = 99;
angle = 99;
end
end
end

function slope = getSlope(v1, v2)
slope = (v2(1) - v1(1)) / (v2(2) - v1(2));
end
```

A.2 MATLAB CODE TO FIND PIXEL POSITIONS

```
function [points, m_result, end_point, bp] = findsegment(m, row,
```

```
        col, bp)
%Find Segment Function
%   Search and detect all possible pixels that are part of a
$segment.
%-----------------------------------------------------------------
%Artifact:  findsegment.m
%Version:   1.0
%Date:      15/03/2020 12:35:00
%Author:    Miguel Angel Gil Rios
%Email:     angel.grios@gmail.com
%-----------------------------------------------------------------
%Usage:
%   [points, m_result, end_point] = findsegment(m, row,
%   col, bp)
%
%Inputs:
%   m:   A Logical matrix.
% row:   The row index where the start point pixel is located.
% col:   The column index where the start point pixel is located.
% bp:    An optional array of Nx2 indicating the current
%        found branch points locations (row, col).
%
%Outputs:
%        points:A Nx2 matrix containing the positions (row,
$        column) with pixels that are part of the segment.
%     m _result:A copy of the m input matrix with zeros in the
%        positions that were identified as part of the segment.
%        end_point:A 1x2 vector containing the position (row,
%        col) of the segment end reached by the function.
%        Segment end can be reached under next conditions:
%        1.  No more ways to explore are available
%            (all neighborgs are 0).
%        2.  We fall in a branch point from which a non unique
%            path is possible to follow.
%        3.  We reach some of the matrix boundaries and
%            there is no more remaining positions to explore.
%            However, you can assume additional considerations:
%        1.  If row and col values are similar to those stored
%            in end_point there is probably to exists
%            only an isolated pixel but not a segment.
%        2.  You can check if end_point values are similar to
%            another end-point that you are keeping in your
%            computing code and you could consider to remove it.
%
```

```
%------------------------------------------------------------------
points = zeros(0);
m_result = m;
end_point(1, 1:2) = -1;

row_current = row;
col_current = col;

k = 0;

flag = true;

if ~exist('bp')
bp = zeros(0);
end

%We reach the end of the segment if (flag == false) under next
%conditions:
%    1. No more ways to explore are available(all neighborgs
%  are 0).
%    2. We fall in a branch point from which a non unique path
%        is possible to follow.
%    3. We reach some of the matrix boundaries and there is no
%        more remaining positions to explore.
first_time = true;
while flag == true
k = k + 1;
points(k, 1) = row_current;
points(k, 2) = col_current;

%The last point (end-point) explored is (row_current,
%                                        col_current):
end_point(1, 1) = row_current;
end_point(1, 2) = col_current;

%Extract a 3x3 window centered at (row_current,
%                                  col_current):
sw = eswm(m_result, row_current, col_current, 3, 3, true);
sw(2, 2) = 0;
sum_pos = sum(sw(:));
if sum_pos == 0 %First flag case
if row ~= row_current || col ~= col_current
if frindex(points, [row_current col_current]) == 0
points(k + 1, 1) = row_current;
```

```
points(k + 1, 2) = col_current;
end
m_result(row_current, col_current) = 0;
end
flag = false;
else
%Find the positions from 1 to 8 where is posible to
% follow a path:
vpos = findpixdirs(sw);
pos = find(vpos == 1);

%If only one direction is possible
if sum_pos == 1
m_result(row_current, col_current) = 0;

%Check if current position falls in a branch point:
pos_bp = frindex(bp, [row_current, col_current]);
if pos_bp > 0 && ~first_time
flag = false;
else
% Find the next position to explore:
switch pos(1, 1)
case 1
col_current = col_current + 1;
case 2
row_current = row_current - 1;
col_current = col_current + 1;
case 3
row_current = row_current - 1;
case 4
row_current = row_current - 1;
col_current = col_current - 1;
case 5
col_current = col_current - 1;
case 6
row_current = row_current + 1;
col_current = col_current - 1;
case 7
row_current = row_current + 1;
case 8
row_current = row_current + 1;
col_current = col_current + 1;
end
```

```
%Check if new position is out of matrix boundaries:
%Third flag case
flag = row_current > 0 & row_current <= size(m_result,
1) & ... col_current > 0 & col_current <= size(m_result,
2);
end

%Second flag case: we fall in a branch point
else
%Add the position of branch point to the bp array:
if frindex(bp, [row_current col_current]) < 1
bp(size(bp, 1) + 1, :) = [row_current col_current];
end
flag = false;
end
end
first_time = false;
end
end
```

A.3   MATLAB CODE TO EXTRACT A WINDOW FROM A MATRIX

```
function subwindow = eswm(m, ri, ci, h, w, c)
% ESWM  Extract a subwindow from a matrix.
%
%
%    subwindow = eswm(m, ri, ci, h, w, c)
%        Extracts a subwindow of size h*w.
% Params:
% m:  The matrix fom which subwindow will be extracted.
% ri: The row index corresponding to the interest point.
% ci: The column index corresponding to the interest point.
% h:  The number of points that will be taken for the
% subwindow vertically.
% w:  The number of points that will be taken for the
% subwindow horizontally.
% c:  A logical value indicating if the interest point will
%     be located at the center of the sub-window (true) or it
%     will be located at the upper left corner of the
%     sub-window (false).
%     If the intetest point will be located at the center
%     of the sub-window, the h and w values will be scaled
%     to the nearest upper odd value in order to take:
%      floor(h/2) points above of the interest point
%       floor(h/2) points bellow of the interest point
```

```
%         floor(w/2) points at left of the interest point
%         floor(w/2) points at right of the interest point
%
if ~exist('c','var')
c = false;
end

if c == true
subwindow = eswm_center  (m, ri, ci, h, w);
else
subwindow = eswm_top_left(m, ri, ci, h, w);
end
end

function subwindow = eswm_top_left(m, ri, ci, h, w)
if (ri + h) > size(m, 1) || (ci + w) > size(m, 2)
subwindow = zeros(h, w);
if (ri + h) > size(m, 1)
h = size(m, 1) - ri;
end
if (ci + w) > size(m, 2)
w = size(m, 2) - ci;
end
for i = 1 : h + 1
for j = 1 : w + 1
subwindow(i, j) = m(ri + (i - 1), ci + (j - 1));
end
end
else
subwindow = m((ri : ri + (h - 1)), (ci : ci + (w - 1)));
end
end

function subwindow = eswm_center(m, ri, ci, h, w)
if mod(h, 2) == 0
h = h + 1;
end
if mod(w, 2) == 0
w = w + 1;
end

subwindow = zeros(h, w);

r_ini = ri - floor(h / 2);
```

```matlab
for i = 1 : h
c_ini = ci - floor(w / 2);
for j = 1 : w
if r_ini > 0 && c_ini > 0 && r_ini <= size(m, 1)
&& c_ini <= size(m, 2)
subwindow(i, j) = m(r_ini, c_ini);
end
c_ini = c_ini + 1;
end
r_ini = r_ini + 1;
end
end
```

A.4 MATLAB CODE TO FIND ROW VECTOR

```matlab
function row_index = frindex(m, vector)
%Find Row Index
%    Find the index where the values of a row vector are the
%    same inside a matrix.
%-------------------------------------------------------------
%Artifact:  frindex.m
%Version:   1.0
%Date:      17/03/2020 20:05:00
%Author:    Miguel Angel Gil Rios
%Email:     angel.grios@gmail.com
%-------------------------------------------------------------
%Usage:
%   row_index = frindex(m, vector)
%
%Inputs:
%        m:  A numerical matrix.
%   vector:  The row vector with the values to be found
%            inside the matrix.
%
%Outputs:
%   row_index:  An Integer representing the row number
%               where the vector values were coincident
%               inside the matrix. If no coincident row
%               vector was found, 0 is returned. If there
%               exists more than one coincidence, the first
%               occurence index is returned.
%-------------------------------------------------------------
row_index = 0;
for k = 1 : size(m, 1)
if isequal(m(k, :), vector)
```

```
row_index = k;
k = size(m, 1) + 1;
end
end
end
```

Index

For Product Safety Concerns and Information please contact our EU
representative GPSR@taylorandfrancis.com
Taylor & Francis Verlag GmbH, Kaufingerstraße 24, 80331 München, Germany

www.ingramcontent.com/pod-product-compliance
Lightning Source LLC
Chambersburg PA
CBHW052012230326
41598CB00078B/2861